Innovative Heat Exchanger Technologies, Developments and Applications

Edited by Peixin Dong and Xin Sui

Published in London, United Kingdom

Innovative Heat Exchanger Technologies, Developments and Applications
http://dx.doi.org/10.5772/intechopen.1001647
Edited by Peixin Dong and Xin Sui

Contributors
Alberto Saldaña-Robles, Aragaw Alamnia, Arash Mahdavi, Carlos Alberto González-Rodríguez, Jiahui Ma, Jiawei Li, Jorge Sergio Téllez-Martínez, José Luis Zúñiga-Cerroblanco, Juan Gregorio Hortelano-Capetillo, Julio César González-Juárez, Lingling Tian, Peixin Dong, Seiji Kojima, Shen Tian, Tieying Wang, Xin Sui

First published in London, United Kingdom, 2024 by IntechOpen
IntechOpen is the global imprint of INTECHOPEN LIMITED, registered in England and Wales, registration number: 11086078, 167-169 Great Portland Street, London, W1W 5PF, United Kingdom

British Library Cataloguing-in-Publication Data
A catalogue record for this book is available from the British Library

Additional hard and PDF copies can be obtained from orders@intechopen.com

Innovative Heat Exchanger Technologies, Developments and Applications
Edited by Peixin Dong and Xin Sui
p. cm.
Print ISBN 978-0-85466-461-0
Online ISBN 978-0-85466-460-3
eBook (PDF) ISBN 978-0-85466-462-7

For EU product safety concerns:
IN TECH d.o.o., Prolaz Marije Krucifikse Kozulić 3, 51000 Rijeka, Croatia,
info@intechopen.com or visit our website at intechopen.com.

Meet the editors

Peixin Dong obtained his Ph.D. from the University of Queensland, Australia, and is a researcher at The Hong Kong Polytechnic University. He has published more than thirty academic journal papers on heat transfer, twelve as first author and corresponding author. His research interests include natural draft dry cooling towers, printed circuit heat exchangers, electric vertical take-off and landing (eVTOL), aircraft thermal management, and wave propagations. He has been selected for the Hong Kong ITF-Talent Hub 2024. His mission and dedication are investigating the best solution for the thermal management of eVTOLs, which are expected to relieve urban congestion.

Xin Sui, BS, Ph.D., is an associate professor at the School of Mechatronics Engineering and Key Laboratory of Mechanical Design and Transmission System of Henan Province, Luoyang, Henan, China. He was a visiting academic at the School of Mechanical and Mining Engineering, University of Queensland, Australia. Dr. Xin has published nearly thirty articles as the first author and co-author in international academic journals such as the *International Journal of Heat and Mass Transfer and Applied Thermal Engineering*. His research interests include off-design and transient simulation and performance evaluation of heat exchangers, transient behavior study for sCO2 power block, and machinery condition monitoring and fault diagnosis for rotating machines.

Contents

Preface **IX**

Section 1
Phase Change Materials (PCMs) Heat Exchangers **1**

Chapter 1 **3**
Development of Latent Thermal Energy Storage Heat Exchangers to Utilize Renewable Energies with Thermal Form
by Shen Tian, Jiahui Ma, Lingling Tian, Jiawei Li and Tieying Wang

Chapter 2 **27**
Integration of Phase Change Materials in Advancing Heat Exchangers for Enhanced Utilization of Variable Renewable Energy
by Aragaw Alamnia

Section 2
Modeling Methodologies **47**

Chapter 3 **49**
Recent Progress in Heat Transfer and Fluid Flow Characteristics within Cross-Flow Heat Exchangers: A Passive Approach
by Arash Mahdavi

Chapter 4 **65**
Fundamentals and Semi-Analytical Analysis for Modeling the Capacity of Shell and Tube Heat Exchangers for Temperature Control in Hydrological Applications
by Juan Gregorio Hortelano-Capetillo, Jorge Sergio Téllez-Martínez, José Luis Zúñiga-Cerroblanco, Alberto Saldaña-Robles, Julio César González-Juárez and Carlos Alberto González-Rodríguez

Chapter 5 **103**
Two Modeling Approaches for sCO_2-PCHEs Transients
by Xin Sui and Peixin Dong

Section 3
Material Thermodynamics **115**

Chapter 6 **117**
Dynamic Specific Heat and Glass Transitions
by Seiji Kojima

Preface

The evolving landscape of heat exchanger technology is marked by significant advancements and innovations and is driven by the growing demand for efficient thermal management across various industries. From power generation and chemical processing to HVAC systems and electronic cooling, heat exchangers are integral to numerous industries. This book, *Innovative Heat Exchanger Technologies, Developments and Applications*, provides a thorough exploration of the latest advancements and innovative applications within this critical field.

The book is organized into six comprehensive chapters, bringing together the expertise of leading researchers and practitioners. The first section (two chapters) focuses on phase-change material (PCM) heat exchangers, examining the principles, design considerations, and practical applications of PCMs. This section highlights the potential of PCMs to enhance heat exchanger performance, particularly in energy storage and thermal management solutions.

The subsequent section (three chapters) offers an extensive review of the development of various types of heat exchangers alongside the employed modeling methodologies. This section covers traditional and emerging heat exchanger designs, providing valuable insights into their evolution, characteristics, and applicability across different sectors. These modeling methodologies serve as an essential resource for researchers and engineers aiming to deepen their understanding of heat exchanger technology.

The final section (one chapter) delves into the thermodynamic properties of thermal materials. Understanding the fundamental knowledge is crucial for improving the efficiency of heat transfer systems. Accordingly, this section presents an in-depth example involving heat transfer processes, modeling techniques, and practical considerations.

This book is intended for researchers, engineers, and students engaged in the study and application of heat exchanger technologies. It integrates theoretical insights with real-world applications, narrowing the gap between fundamental research and practical implementation and aiming to inspire further research and boost development in this field.

We extend our heartfelt gratitude to the contributors for their invaluable insights and dedication to advancing heat exchanger technology. We also thank the readers for their interest in this book. We hope that the knowledge shared within these pages will foster continued progress and innovation in the field of heat exchangers.

Peixin Dong
The Hong Kong Polytechnic University,
Hong Kong

Xin Sui
Henan University of Science and Technology,
Henan Province, China

Section 1

Phase Change Materials (PCMs) Heat Exchangers

Chapter 1

Development of Latent Thermal Energy Storage Heat Exchangers to Utilize Renewable Energies with Thermal Form

Shen Tian, Jiahui Ma, Lingling Tian, Jiawei Li and Tieying Wang

Abstract

The latent thermal energy storage (LTES) technology has advantages of high thermal energy storage density, system volume saving, and easy installation, which is of great significance for improving the flexibility of thermal energy supply in the background of renewable energy utilization. The process of storing/releasing thermal energy of LTES systems mainly relies on heat exchangers, one special device for exchanging heat between heat transfer fluids (HTFs) and phase change materials (PCMs). The inherent liquid-solid phase change process of PCMs in the device makes this kind of heat exchanger urgently needed for more efficient design and optimization. In this chapter, working materials, device types, design forms, heat transfer enhancement methods, and some actual applications for heat exchangers are introduced. The present problems of the heat exchanger development are also discussed. The purpose of this chapter is to attempt to help the readers obtain key knowledge and design points for potential better applications of the LTES heat exchangers.

Keywords: phase change material, heat transfer fluid, heat exchanger design, device type, latent thermal energy storage

1. Introduction

Latent thermal energy storage (LTES) is a process of using liquid-solid latent heat to store energy in thermal form (including heat and cold thermal energy) and then release it to the thermal energy systems. This is a typical intermittent and unsteady energy utilization process suitable for balancing and reducing costs for thermal energy usage occasions. In recent years, with the urgent trend of mitigating climate change and promoting green sustainability, the proportion of renewable energy is constantly increasing. The end users have put forward higher requirements for the flexibility of energy supply. Among the many forms of energy usage used by end users, thermal energy is crucial. For instance, in China, thermal energy, which is mostly used for ensuring thermal comfort in buildings (heating and cooling), accounts for 15% of the total social energy consumption. Globally, according to statistics, thermal energy covers 50% of the total energy consumption and is responsible for 40% of the CO_2 emissions. Consequently, it seems that the development of thermal energy storage technology that combines with renewable energy is important for accelerating thermal energy utilization toward green sustainability. Based on this, it is forecast by the International Renewable Energy Agency (IRENA) that the global thermal energy storage market will triple in size by 2030, and the capacity of thermal storage devices will increase to over 800 GWh in the next 10 years [1]. Therefore, in order to achieve an approximate effect of direct thermal energy usage for the end users, the LTES technology has received widespread attention. The scenarios of LTES applications can be divided according to the required temperature range, which is shown in **Figure 1**. It can also indicate that this technology has a wide range of application demands and applicability.

The storage and release of thermal energy in LTES devices mainly rely on the heat transfer process between the mediums. Therefore, these devices can be considered as a special kind of heat exchanger. The design, optimization, and control methods for traditional heat exchangers are also applicable to the LTES heat exchangers. The difference is that they are composed of PCMs, encapsulations, and HTFs. Consider for a moment the heat transfer process within a shell-and-tube LTES heat exchanger, such as that shown in **Figure 2(c)**. The HTF flows in the tube to pass heat through the wall of the tube. Subsequently, the PCM inside the shell is frozen or melted. The liquid-solid interface of PCM dynamically and continuously changes along with the tube

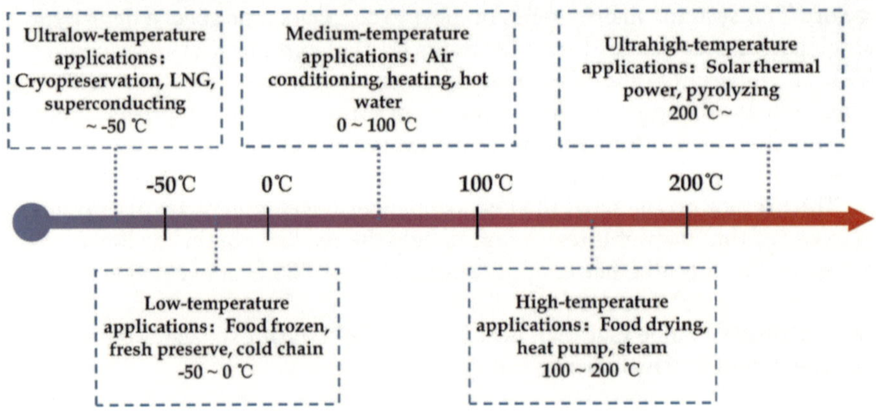

Figure 1.
Application scenarios and temperature ranges of LTES devices.

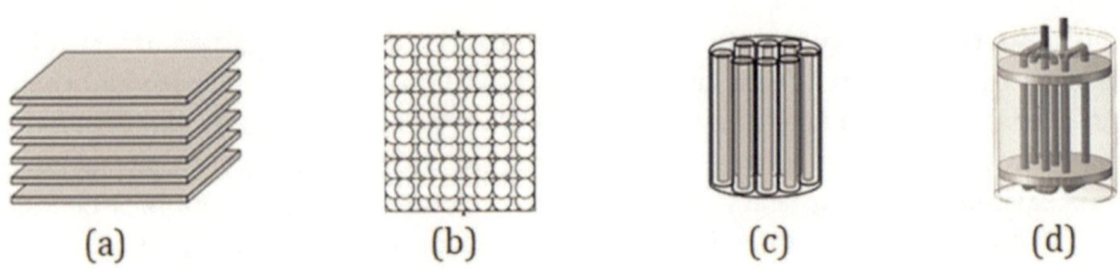

Figure 2.
Basic forms and characteristics of LTES heat exchangers: (a) plate type; (b) sphere type; (c) tank type; (d) casing tube type.

path. Yet the problem of determining the heat transfer in this situation, however difficult it appears to be, is a task that must be undertaken. The HTF flow in the tubes seems more tractable, but it, too, brings with it several problems that are not as simple as flows in a tube with a flat internal surface for the effect of the phase change process outside the tube. This kind of heat exchanger, similarly to the others, presents a kind of microcosm of internal and external heat transfer problems. These issues extend to the applications mainly reflected in the non-uniformity of the working process, such as:

a. The longest duration in the final stage of frozen or melting process;

b. The low efficiency for the overall heat transfer process due to low thermal conductivity of PCM;

c. Thermal stratification and phase separation caused by the poor fluidity of the liquid PCM.

The above are also urgent issues that need to be addressed for the research and development of this type of heat exchanger.

Our aim in this chapter is to present the analysis of a few basic concepts and to show progression toward increasing applications for LTES heat exchangers. We begin this undertaking with the fundamental problem of the materials.

2. Materials

The LTES heat exchanger consists of PCM, encapsulation containment, and HTF. Generally, in order to ensure the stability of PCM for cycling, the PCM is confined in encapsulation containment from several milliliters to several liters. The HTF flows within fluid channels that are composed of the outer side of the containments. The main material characteristics and selection criteria of these different functional parts in the heat exchangers are described below.

2.1 Phase change material

The term PCM specifically refers to materials that can produce liquid-solid phase transition in heat transfer process. The temperature at which the phase change happens is called the phase change temperature (PCT). The selection of the PCM is an essential step at the initial design stage of the LTES heat exchanger, which determines the storage capacity and eventually affects the device's final application. It is widely accepted that the desired requirements for PCMs are:

a. An appropriate operating temperature range depending on the application;

b. High energy density;

c. High thermal conductivity;

d. Low degree of subcooling;

e. Corrosion resistance;

f. Chemical and physical stability;

g. Not poisonous, toxic, flammable, or explosive;

h. Small vapor pressure and volume change;

i. Low cost and available in large quantities.

Based on the above criteria, many kinds of pure materials have been selected and applied. According to their chemical nature, the materials can be classified into organic and inorganic substances with variable PCTs. Then, one can deduce that more than two pure substances can be mixed for the PCT adjustment by varying the content of the components, which often refers to a mixture of PCMs. Among them, the eutectic PCM is fabricated from more than two substances. It can utilize hydration to store or release thermal energy, which has advantages in terms of higher energy storage density. In addition, these pure or mixed PCMs have some natural limitations for widespread applications, mainly due to their corrosiveness, supercooling, and phase segregation. Innovative nano additive and composite PCMs have been developed, which have given the ability to decrease subcooling, nucleation time, thermal conductivity improvement, and leakage prevention.

Based on temperature ranges as mentioned in the above section, the commonly used PCMs are listed and discussed below.

2.1.1 PCMs for low-temperature applications

When the PCT is below 0°C, it can be considered that this type of PCM is used for cold thermal energy storage. They are widely applied in areas such as frozen food, fresh conserve, cold chain, and air conditioning during the summer season. Compared with other applied PCMs, these low-PCT PCMs are mainly composed of pure substances and simple solutions, which also make them very easy to be used. The commonly seen PCMs for low-temperature applications are listed in **Table 1**. The PCMs that correspond to the ultra-low-temperature range are mainly composed of volatile pure substances. In temperature range within −50 to 0°C, many types, including pure substances, metallic salts solutions and eutectic water-salt solutions, can be utilized. It may be worth noting that water has the largest latent thermal energy density compared to other PCMs [2]. The additives and components, in general, can reduce the latent heat, increase the cost of the PCM products, and deteriorate the cycling stability. Therefore, the effect of including components should be evaluated to ensure an overall enhancement in the performance of PCMs. For instance, the enhancement may focus more on PCT fabrication and thermal conductivity improvement.

2.1.2 PCMs for medium-temperature applications

When the PCT is distributed between 0 and 100°C, it can be considered that this type of PCMs is used for heat thermal energy storage. This temperature range usually corresponds to the range of atmospheric liquid water applications. Therefore, from the supply-side view point, the PCMs are suitable for storing low-grade waste heat, solar energy, and geothermal energy. In addition, at the end-use view point, they are

Material	PCT (°C)	Thermal conductivity ($W\ m^{-1}\ K^{-1}$)	Latent heat ($J\ g^{-1}$)
Nitrogen (N_2)	−209.85	0.163 (l)	25.80
Pentane (C_5H_{12})	−129.75	0.173 (l)	116.43
Methanol (CH_4O)	−97.15	0.210 (l)	99.25
Methyl ethyl ketone (C_4H_8O)	−86.65	0.170 (l)	117.05
Hexanone ($C_6H_{12}O$)	−55.45	0.160 (l)	148.70
Ethylene glycol/H_2O (50:50 wt%)	−37.10		41.59
39.6 wt% K_2CO_3	−36.50		165.36
22.4 wt% NaCl	−21.20	0.570 (l)	222–235
Diethylene glycol ($C_4H_{10}O_3$)	−10.40	0.200 (l)	247.00
n-Dodecane ($C_{12}H_{26}$)	−9.50	2.210 (l)	216.20
27.2 wt% $ZnSO_4$	−6.50		208–235.75
Caproic acid ($C_6H_{12}O_2$)	−4.00	0.150 (l)	146.18
Water (H_2O)	0.00	0.551 (l)	333.30

Table 1.
Properties of PCMs for low-temperature applications.

Material	PCT (°C)	Thermal conductivity ($W\ m^{-1}\ K^{-1}$)	Latent heat ($J\ g^{-1}$)
Formic acid	8	0.27 (l)	277
Acetic acid	17	0.19 (l)	192
Calcium chloride hexahydrate	30	0.53 (l)	125
Lauric acid	44	0.15 (l)	212
Stearic acid	54	0.17 (l)	157
Sodium acetate trihydrate	58	0.34 (l)	266
Palmitic acid	61	0.17 (l)	222
Paraffin wax	0–90		150–250
Acetamide	82	0.25 (l)	260
Magnesium nitrate hexahydrate	89	0.50 (l)	140

Table 2.
Properties of PCMs for medium-temperature applications.

suitable for releasing thermal energy applications such as household heating, automotive air conditioning, and battery cooling. The commonly seen PCMs for medium-temperature applications are shown in **Table 2**.

It can be seen in **Table 2** that the organic PCMs, which are fatty acids, sugar alcohols, carboxylic acids, amides, and alkanes, are mainly employed for medium-temperature applications. Compared with inorganic PCMs, they have attracted more attention from the perspective of real applications due to their advantageous properties such as non-toxicity, slight supercooling, and good cycling stability. However, the character for having carbon atoms in their structure generally has very low thermal conductivity (from 0.1 to 0.6 W m^{-1} K^{-1}), hence requiring new methods to enhance their heat transfer performance. Therefore, nanoparticles, nanosheets, nanofibers, nanotubes, nanowires, and nanorods have been added and evaluated for the improvement of thermal conductivity [3].

2.1.3 PCMs for high-temperature applications

The PCM with a PCT above 100°C, which corresponds to hot steam conditions, can be regarded as for high-temperature applications. This kind of PCM is widely used in areas of concentrating solar power and integrated thermophotovoltaic systems. With these PCMs, solar energy can be harvested and stored in the systems to be efficiently used at night for power generation. It has been reported that a heat-to-electricity efficiency greater than 50% can be achieved. The commonly used PCMs for high-temperature applications are shown in **Table 3**.

It can be seen that the PCMs corresponding to high-temperature applications are mainly composed of organic compounds (sugar alcohols <200°C), molten salts (>300°C), and metallic alloys (>500°C). Molten salt-based PCMs are usually low-cost and have high heat-storage density, although the major disadvantage is the corrosion of containers, pipes, and valves. In addition, there are carbonate-based PCMs can be applied at higher temperatures (around 1000°C) applications. They have the advantages of high melting temperature, high viscosity, and high latent heat, making them suitable for developing structured PCMs in high-temperature thermal storage areas.

Material	PCT (°C)	Latent heat (J g^{-1})
Erythritol	117	340
$LiNO_3$(34)-66KNO_3	133	150
Maleic acid	141	385
$LiNO_3$(58)-42KCl	160	272
$LiNO_3$(87)-13NaCl	208	369
$NaNO_3$(86)-14NaOH	250	160
Zn(52)-48Mg	340	180
NaCl(56)-44$MgCl_2$	430	320
$CaCl_2$(52.8)-47.2NaCl	500	239
KCl(45)-55KF	605	407
LiF(70)-30MgF_2	728	520
NaF(68)-32CaF_2	810	600
Si(56)–44Mg	946	757

Table 3.
Properties of PCMs for high-temperature applications.

2.2 Materials for heat transfer fluid

The HTF is used to deliver heat to the PCMs. In most cases, the reason for choosing fluids as HTFs is their excellent heat transfer performance compared to other mediums and controllability. The HTFs can be divided into three basic types based on their working modes: pure gas phase, gas-liquid phase, and pure liquid phase. For low-temperature applications, the gas-liquid phase flow of volatile media is mainly used; For medium-temperature applications, pure liquid phase or pure gas phase can be used. For high-temperature applications, it is common to use high-temperature and high-pressure gases, which also include gas-liquid phase transitions of the hot steam. The heat transfer performance of the HTFs is usually better than that of the PCM. Therefore, on the HTF side, the significance of heat transfer enhancement is not very prominent. However, the uniformity of the HTF flow and the input power consumption need to be considered in the design process due to the diversity and complexity of the flow channels within the devices. In addition, due to the connecting effect of the HTF between systems and the LTES devices, from a system perspective, the availability and convenience of the HTF material also need to be considered for system optimization. The commonly used HTFs are mainly refrigerants, water, water-based solutions, hot steam, and gases.

2.3 Materials for encapsulation containment

Encapsulation containment refers to confining PCMs in closed containers from several milliliters to several liters. It can be used as a standalone LTES unit or as a multiple-stacked LTES device. The encapsulations can be divided into microencapsulation, microencapsulation, and porous stable shaping. The microencapsulation is the most widely used PCM encapsulation technique. It packages large amount of PCM in a big size and protects the PCM from contamination, which can also help to enhance heat transfer by providing a large heat transfer surface and reducing thermal resistance due to limited thickness. The microencapsulation and the porous stable shaped techniques, on the other hand, create a larger surface area than the microencapsulation, improving the heat transfer, the mechanical stability, and the cycling stability for the actual applications. Materials for the encapsulation containment can be metallic, polymer, and ceramic. The key property of encapsulation containment is its mechanical stability [4]. However, other properties such as density, volatile organic compound emission, flammability, and permeability of the containment shell play a role when a specific application is considered. In the case of micro and nanocapsules, capsule size distribution and morphology, which affect the heat transfer performance of the whole device, also need to be considered.

3. Device types and variations

The structure types of LTES heat exchangers are usually divided into plate type, spherical stacking bed type, tank type, and casing tube type. In order to improve heat storage efficiency, variations from the basic types have also been developed. The details on the device types and the variations are as follows.

3.1 Basic types and features

Plate type: The geometry of the plate is cuboid; as shown in **Figure 2(a)**, the heat transfer fluid (HTF) flows through the plate for heat exchange. The advantages are small pressure drop, simple structure, easy to manufacture and low price. The HTF is located outside of the containment and can flow freely.

Spherical stacking bed type: The basic unit of the spherical stacking bed type is the spherical capsule, as shown in **Figure 2(b)**. The PCM is encapsulated in a spherical shell with a certain thickness and then placed in a storage tank. The HTF exchanges heat through the voids between spheres. The dispersed mode of the spheres has the advantages of a large unit heat transfer area, simple structure, and no leakage. The stacking methods of the spheres can be classified into structured and unstructured. Similar to the plate form, the HTF is located outside the package and can flow freely.

Tank type: This type of device adds the pipes to the energy storage tank, where the PCM is filled in the tank. As shown in **Figure 2(c)**, the HTF flows in the pipe and exchanges heat with the PCM. The number and diameter of pipes in the tank greatly influence energy storage efficiency. The pipes are arranged horizontally or vertically. The HTF is located in the package, and the flow area is limited.

Casing tube type: The advantages of the casing tube include its compact structure and its large heat transfer area, as shown in **Figure 2(d)**. The space of the outer wall surface of the PCM is limited by an outer wall surface. The casing structure can be divided into horizontal and vertical structures according to the placement direction in a tank, and the body radius is the main factor affecting its performance. The HTF is located in the packaging and has a limited circulation area.

In the viewpoint from the surface structure of heat transfer interface (HTI) on the PCM side, the HTI can be divided into three different classifications as shown in **Figure 3**, i.e., flat surface type without surface microstructures, extended surface type with additional areas, and direct contact type that without solid walls as shown in **Figure 3(c)**. In detail, the flat surface HTI exhibits weaker overall contact capability due to its small contact area. The contact capacity of the extended HTI is higher than that of the flat surface HTI. For the direct contact HTI, the contact capacity performs best for the deformed contact interface.

The flat surface HTI comprises a flat surface interface that can be formed for the encapsulation of PCM with plates, spheres, tanks, and casing tubes. As the flat surface HTI does not include small local interface structures, the overall convective heat transfer coefficient can be calculated with respect to the total area of the HTI at the PCM side:

$$U = \frac{\varphi}{A \cdot \Delta T} \tag{1}$$

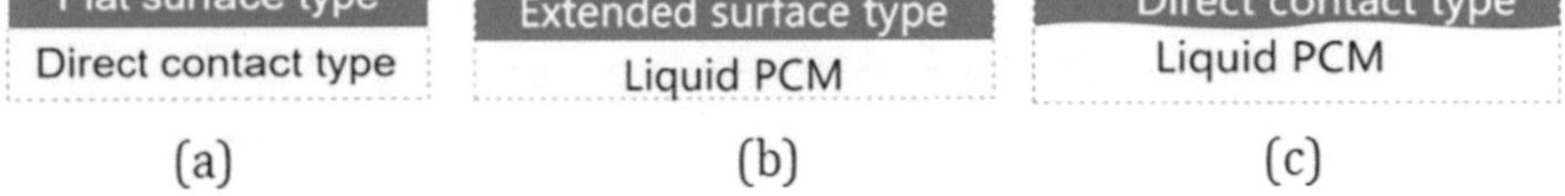

Figure 3.
Classifications of the HTI: (a) flat surface type; (b) extended surface type; (c) direct contact type.

$$\Delta T = \frac{\Delta T' - \Delta T''}{\ln\left(\frac{\Delta T'}{\Delta T''}\right)} \tag{2}$$

$$\varphi = \underset{HTF}{\dot{m}} C_{HTF}(T_{in} - T_{out}) \tag{3}$$

Eq. (1) represents a general calculation formula for the overall heat transfer process of all kinds of heat exchangers. Furthermore, this equation can describe the heat transfer process from the perspective of time averaging. Eq. (2) calculates the logarithmic mean temperature difference of the overall heat exchange process. The inlet temperature of HTF in Eq. (2) is usually maintained constant for the actual applications. The outlet temperature of the HTF can be calculated by Eq. (3). Generally, the temperature change at the outlet is small during the heat transfer process; therefore, a constant heat transfer flow rate is assumed [5]. By using Eqs. (1)–(3), the heat transfer performance of the LTES heat exchanger with different types of HTI can be compared.

3.2 Variations of the device structure

In order to strengthen the heat transfer for the LTES heat exchanger, the improvement of contact ability between the PCM and the HTI is usually adopted. It can be derived from two basic variations, i.e., HTI extension and direct contact. The extended surface HTI improves the heat transfer process by increasing the contact area of the flat HTI. Differently, the direct contact method strengthens the heat transfer process through the deformable direct contact HTI between PCM and HTF.

3.2.1 The extended surface HTI

The surface extension of the HTI is a widely applicable heat transfer improvement method for almost all of the basic types of LTES heat exchangers. From a general viewpoint, for heat exchangers, the evaluation of the extended surface heat exchangers involves the use of the convective heat transfer coefficient and Nusselt number ratio [6]. However, the Nusselt number of the LTES heat exchangers, owing to the small thermal resistance of HTI and a variable thermal resistance on the PCM side, cannot be precisely calculated. Therefore, the ratio of the overall heat transfer coefficient is used to clarify the enhancement performance in this section. The corresponding equations are listed below:

$$\frac{U}{U_0} = \frac{Q \cdot A_{PCM0} \cdot \Delta T_0}{Q_0 \cdot A_{PCM} \Delta T} \tag{4}$$

$$U = \frac{1}{\frac{1}{h_{HTF}} + \frac{\delta_{wall}}{\lambda_{wall}} + R_{PCM}} \tag{5}$$

3.2.2 Direct contact HTI

The term "direct contact" refers to exchange heat between one medium and another material directly. In other words, for the heat transfer process, the PCM and the HTF share the same HTI, and the HTI often shows a dispersed style (such as droplets and bubbles). The physical property variations through the HTI are discontinuous. Meanwhile, the HTI shows obvious instantaneous deformation and complex structural characteristics. Therefore, the area of the HTI is difficult to obtain precisely. However, the overall convective heat transfer coefficient cannot be directly calculated without the precise determination of the heat transfer area. Based on this, researchers developed an estimation method that evaluates the mean size of the contact area. The details are described as follows.

First, the Galileo and the Reynolds numbers are calculated by using the physical parameters of PCMs and HTF:

$$\mathrm{Ga} = \frac{d^3 \cdot \rho \cdot (\rho_d - \rho_{PCM}) \cdot g}{\mu^2} \tag{6}$$

$$Re = \left(2.33 \cdot Ga^{0.018} - 1.53 \cdot Ga^{-0.016}\right)^{13.3} \tag{7}$$

Second, due to the interference of the dispersed phases in the direct contact type, the Prandtl number is introduced. Then, the Nu that represents the ratio of convective heat to conduction heat can be obtained by Eq. (9):

$$\mathrm{Pr} = \frac{\mu \cdot C_{PCM}}{\lambda} \tag{8}$$

$$Nu = 2 + 0.65 \cdot \mathrm{Re}^{0.5} \cdot \mathrm{Pr}^{\frac{1}{3}} \tag{9}$$

The overall convective heat transfer coefficient can then be obtained as follows:

$$Nu = \frac{U \cdot d}{\lambda} \tag{10}$$

The diameter of the dispersed phases must be specified when calculating the Eq. (10). The diameter values can be denoted and predicted from the literatures for the calculations. What is more, in terms of evaluating the accessibility for direct contact type, calculation of the surface area density is also needed. It can be obtained by estimating the mean diameter values. In addition, the below equations can also be used to obtain the surface area by estimating the heat transfer coefficient for the direct contact HTI:

$$\varepsilon = \frac{T_{\mathrm{in}} - T_{out}}{T_{\mathrm{in}} - T_{PCM}} \tag{11}$$

$$\varepsilon = 1 - \mathrm{e}^{-NTU} \tag{12}$$

$$NTU = \frac{UA}{\dot{m}C} \tag{13}$$

3.3 Heat transfer performance of LTES heat exchangers

3.3.1 Evaluation and analysis of the flat HTI type

Usually, we can compare and analyze the heat transfer performance of the LTES heat exchangers by using parameters such as the overall convective heat transfer coefficient, surface area density, and heat transfer efficiency. The essential difference between the four basic types is the accessibility between PCM and HTF [7]. Thus, the surface area density represented by the area-to-volume ratio is used for the quantification of accessibility. A literature review of the data and the calculations of LTES heat exchangers with flat HTI is conducted in this section. The results of heat transfer coefficient and surface area density are extracted to form the upcoming **Figure 4**, which gives a direct look into the heat transfer performance with different LTES heat exchangers.

It can be seen in **Figure 4** that the overall convective heat transfer coefficient almost linearly increases with the increase in surface area density for all the cases. From comparisons between different device forms, the same variation trend can also be observed. The plate from the LTES heat exchanger comprises different numbers of parallel plates. The HTF flows between the plates. However, the plate form usually has a large aspect ratio, which exhibits a slight discreteness along the length direction. In addition, this form is usually applied when air is used as the HTF and the air flows in an interpenetration channel. Therefore, the overall convective heat transfer coefficient and the surface area density of the plate form are distributed in the lower left zone in **Figure 4**.

For the spherical stacking bed form, the size and the filling approach of the spherical capsule are the main factors affecting the energy storage efficiency. Many researchers have conducted experiments and simulations to prove this. The results indicate that when the volume of the tank keeps the same, the overall convective heat transfer coefficient of the heat exchanger is limited by the size and the number of filled spheres. The reason for this phenomenon is that the surface area density of the filled spheres has varied according to size and number. Furthermore, the nonstructural distributions (without arranged positions for the spheres) with the same number and size of the spheres can provide higher overall heat transfer coefficients than the structural distributions. The reason is that the nonstructural is more dispersed than the structural distributions. Overall, the spherical stacking bed form is conducive to the dispersed heat transfer process of the PCM and can better improve the heat transfer performance. Therefore, in **Figure 4**, it has a larger heat transfer performance than the plate form.

For the tank form, the number of pipes, the pipe layout direction, the pipe diameter, and the pipe outlet position are the main factors affecting the energy storage process. Researchers also conclude that the larger the pipe diameter, the better the heat transfer performance can be obtained. In **Figure 4**, the overall convective heat transfer coefficient of the tank form is generally higher than that of the spherical

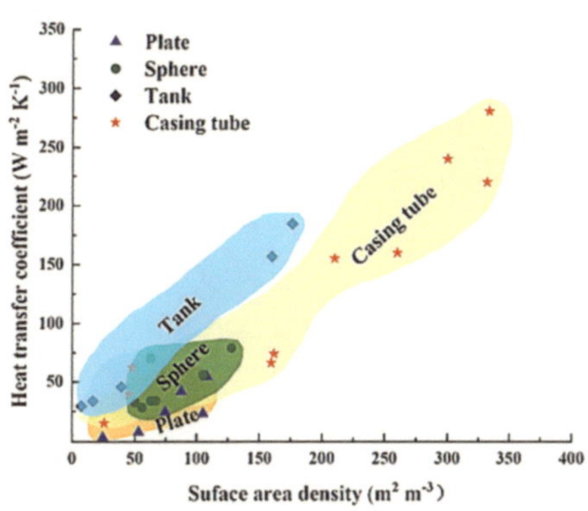

Figure 4.
Surface area density and corresponding overall convective heat transfer coefficient for different LTES heat exchangers with flat HTIs.

stacking bed type. This is because the HTF in the tank form is constrained inside the PCM by the pipe wall, which increases the accessibility of the PCM and the HTF. Therefore, inspired by this, it is more beneficial to improve the heat transfer performance by adding the HTI into the PCM.

Similarly, the HTF of the casing tube form is limited inside the PCM by pipe walls. The highest corresponding heat transfer performance can be seen in **Figure 4**. Unlike the tank form, the PCM in casing tube form is also limited by an outer shell wall. According to the literature review, the radius ratio of the shell to tube is usually smaller than the tank form, which provides a higher surface area density than the other forms.

3.3.2 Evaluation and analysis of surface extension HTI type

It is a widely used enhancement method to expand the heat transfer area on the PCM side for the LTES heat exchanger improvement. Based on the development principle of the LTES heat exchangers, it is found that the most common methods for surface extension are adding fins, embedding heat pipes, and adopting gradient-sized surfaces. To distinguish the performances before and after extending the surfaces, the ratio of surface area density is adopted. This value represents the extent strength to which the LTES devices can be improved after strengthening. Similarly, the heat transfer coefficient ratio is also used to represent the heat transfer enhancement. In addition, considering the evaluation of the heat transfer enhancement effect, the only difference between the tank and casing-tube structures is the outer shell wall. Both heat transfer enhancement methods are mainly aimed at the expansion of the outer wall of the inner tube. Therefore, the casing tube and tank forms are combined in the analysis; the comparison results are shown in **Figure 5**.

3.3.3 Evaluation and analysis of direct contact HTI type

The LTES heat exchangers with direct contact HTI type can be divided into two forms, top-down and bottom-up, based on the distributions of the HTF inlet. The HTF that is usually used is pure gas and pure liquid. It should be noted that the phase change (gas-liquid transition) of HTF can also significantly enhance the heat transfer

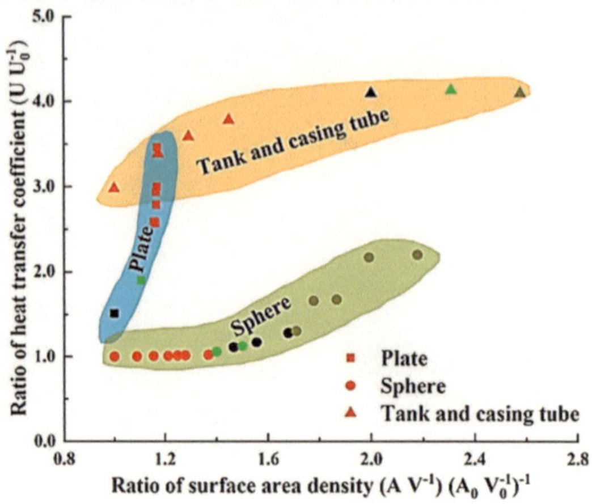

Figure 5.
Surface area density ratio and heat transfer coefficient ratio of different forms with surface extension HTIs.

performance for this type of heat exchanger. The direct contact HTI is formed by two or more immiscible media. Its structure and position are highly variable and are driven by the natural or forced flow of the HTF. Since the variation of HTI can highly improve the mobility of the liquid PCM, this type of heat exchanger has higher heat transfer performance than the others. The estimation of the average size of the HTI seems crucial for calculating the total convective heat transfer coefficient. However, few studies can accurately predict this value for the inherent complex mechanism of the HTI variation. Further studies are needed to quantitatively evaluate the heat transfer properties of the direct contact LTES heat exchangers. Another method for the evaluation of heat transfer performance is direct experiments, which are compared to indirect types. Researchers concluded that the maximum increase in the average rate of energy storage for direct contact type is 218%. Due to the removal of the fixed wall surface, the heat transfer coefficient at the interface between PCM and HTF can reach higher levels, beyond 1000 W m^{-2} K^{-1}.

4. Design methods

Usually, constructing models is a crucial step for the heat exchanger design. The developed models for heat exchangers can be divided into one-dimensional steady models, dynamic analytical models, and numerical models, which play indispensable roles in the heat transfer area. The modeling process for heat transfer issues is not only a component of scientific research but also a cornerstone in solving practical problems in engineering practice. For the LTES heat exchangers, the design also relies on the development of the heat transfer model. Therefore, in this section, the design methods are described by developing the specific models, which can also provide the guidelines for the actual device design.

4.1 One-dimensional steady model

The study of system behavior under equilibrium or stable conditions is crucial for the ease of implementation of the LTES heat exchanger design. Common types of one-dimensional models include static steady-state models, which mainly focus on the equilibrium state at specific time points. It is worth noting that static steady-state models are usually simpler, but provide valuable insights into system behavior.

4.1.1 Construction of one-dimensional model for LTES heat exchangers

When constructing steady-state models, it is essential to ensure that the model's parameters and assumptions align with real conditions. Conducting sensitivity analysis is crucial to understanding the model's response to changes in parameters. Furthermore, the significance of the one-dimensional model lies in its ability to serve as a references for the long-term performance of systems, providing a foundation for strategy formulation and decision-making. Therefore, in practical applications, model results should be compared with actual observational data to validate their accuracy and reliability.

4.1.2 Modeling of equations

In a one-dimensional model, equations typically articulate the equilibrium conditions among various variables in a system under steady conditions. With a relatively fixed heat transfer area, the state parameters encompass the temperature variation of PCM and the cold/hot supply performance.

Its mathematical representation is based on the mean effectiveness-NTU (number of transfer units) method. When the working fluid within the heat exchanger undergoes a phase change, the effectiveness of the heat exchanger can be expressed as follows [8]:

$$\varepsilon = 1 - e^{-NTU} \tag{14}$$

When utilizing PCM, effectiveness is formulated as the ratio of the actual released heat energy to the theoretical maximum released heat energy. The NTU is associated with the properties of the working fluid without phase change and is determined by the mass flow rate, heat capacity, and total thermal resistance. The NTU at any given time can be expressed as follows:

$$NTU = \frac{UA}{\dot{m}_h C_h} = \frac{1}{R_T \dot{m}_h C_h} \tag{15}$$

Taking the example from the authors' research of a latent heat energy storage panel assisted by embedded heat pipes (HP), the total thermal resistance (R_T) is a function of six thermal resistances shown in **Figure 6** and defined by Eq. (16):

$$R_{Total} = R_f + R_{c1} + \mathrm{R}_{ew} + R_{cw} + R_{c2} + \mathrm{R}_{pcm} \tag{16}$$

The following equations shown in **Table 4** are used for the calculation of the total thermal resistance. It should be noted that under steady-state conditions, the total thermal resistance of the HPs is formed by the thermal resistance of the HP wall and the wick. The thermal resistance of the vapor flow can be neglected.

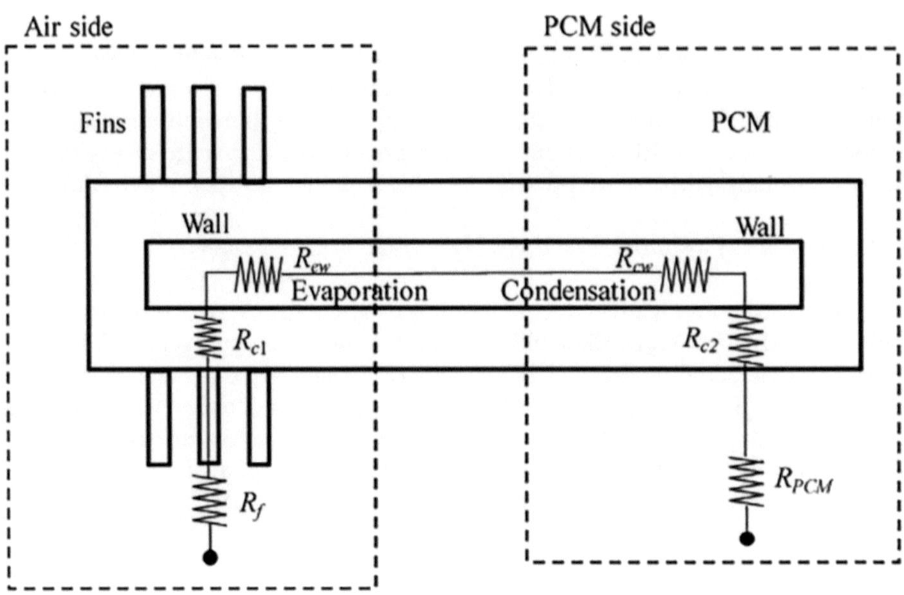

Figure 6.
Thermal resistance model of the heat exchange process.

Thermal resistance	Equations
R_{c1}	$R_{c1} = \ln(r_o/r_i)/(2\pi k_{wall} L_e)$
R_{ew}	$R_{ew} = \ln(r_i/r_v)/(2\pi L_e K_{eq})$
R_{cw}	$R_{ew} = \ln(r_i/r_v)/(2\pi L_c K_{eq})$
R_{c2}	$R_{c1} = \ln(r_o/r_i)/(2\pi k_{wall} L_c)$
R_f	$R_{c1} = 1/\left(A_f \eta_f \eta_n h_f\right)$
R_{PCM}	$R_{PCM} = \ln(r_m/r_o)/(2\pi k_{PCM} L_c)$

Table 4.
Referenced equations for the calculation of thermal resistances.

From these equations, it can be observed that the effectiveness varies with the change in the radius of the molten PCM layer. Thus, the average effectiveness of an individual unit can be defined as:

$$\overline{\varepsilon} = \int_{r_m=r_0}^{r_m=r_{\max}} \frac{\varepsilon dr_m}{r_{\max} - r_0} \tag{17}$$

Subsequently, the average performance of cold and hot energy supply can be obtained:

$$\overline{q} = \overline{\varepsilon}\dot{m}c_p(T_{\text{inlet}}\text{-}T_{\text{PCM}}) \tag{18}$$

4.2 Dynamic analytical model

The application of dynamic models in the design is aimed at a more comprehensive understanding of the heat exchangers dynamically. In contrast to the static models that focus on the system's equilibrium state at a specific moment, dynamic models, by modeling along the time dimension, can capture the dynamic characteristics of the device over time, providing explanatory power for both long-term trends and instantaneous changes. Common dynamic modeling approaches usually involve the development of differential and integral equations. The differential equations focus on the continuous-time evolution of the heat transfer process. The integral equations involve the accumulation of system states. Dynamic models can also be categorized into linear and nonlinear models. The linear dynamic models assume that the system's behavior can be described through linear relationships, whereas the nonlinear dynamic models offer more flexibility in handling nonlinear effects within the system, providing a closer representation of real scenarios.

4.2.1 Construction of dynamic analytical model

When constructing dynamic analytical models, special attention must be given to the selection of model parameters and the establishment of initial conditions, as they significantly impact the predictions. Furthermore, conducting analyses on the stability and sensitivity of the model is a crucial step to ensure that the results possess reasonable explanations under varying conditions.

4.2.2 Modeling of equations

In dynamic analytical models, it is essential to initially establish the system's state variables and the relationships between these variables. This can be accomplished through the formulation of differential equations and calculus expressions. In this section, an example of LTES heat exchanger development using bubble flow from the author's research is used to explain the modeling method. When dealing with phase transition processes characterized by unfixed heat transfer interfaces, the utilization of dynamic analytical models can facilitate the swift derivation of simulation outcomes.

In the mathematical representation of dynamic analytical models, the differential equation of energy conversion is [9]:

$$\rho_{Ps}\gamma \frac{dV_{Ps}}{dt} + \rho_{Ps}C_{Ps}\frac{dV_{Ps}}{dt}\Delta T = \dot{m}_h C_h (T_{in} - T_{out}) \tag{19}$$

The calculation of ΔT adopts the logarithmic mean temperature difference method, expressed as:

$$\Delta T = \frac{T_{in} - T_{out}}{\ln\left(\frac{T_{in} - T_P}{T_{out} - T_P}\right)} \tag{20}$$

Consistent with the method of calculating NTU using steady-state models, the NTU at any given time can be represented as follows:

$$NTU = \frac{UA}{\dot{m}_h C_h} = \frac{1}{R_T \dot{m}_h C_h} \tag{21}$$

Taking the direct contact phase change process as an example, the intricate and diverse nature of the contact interface structure necessitates the establishment of state parameters from an overall process perspective. The state variables encompass changes in the volume of the PCM and the outlet temperature throughout the phase transition. Simultaneously, in order to solve this model, certain objective and subjective assumptions have been proposed.

The dimensionless relationship between volume and contact area with HTF during PCM phase transition is assumed as:

$$\frac{dA}{dA \cdot A_0} = \frac{dV}{dt \cdot V_0} \tag{22}$$

$$UA_0 = \frac{1}{R_{T0}} \tag{23}$$

Connect volume and thermal resistance based on the volume assumption can be obtained by:

$$\frac{1}{R_T} = \frac{V}{V_0} \cdot \frac{1}{R_{T0}} \tag{24}$$

To derive the mathematical expression for NTU, Eq. (24) is substituted into Eq. (19), facilitating a dynamic analysis of the heat transfer process in the direct contact LTES heat exchanger:

$$NTU = \frac{1}{\dot{m}_h C_h} \cdot \frac{1}{R_{T_0}} \cdot \frac{V}{V_0} \tag{25}$$

Once the solution to the dynamic analysis model is obtained, it can be applied to predict the future behavior of the system or analyze its response under different conditions. This aids in a comprehensive understanding of the system's dynamic characteristics, offering guidance for optimization in design and control.

4.3 Numerical model

The application of numerical models in the fields of heat exchanger development aims to address complex mathematical equations or encountered practical problems, which are often too difficult to be solved by the analytical models. The major development methods of the numerical models include the finite element method, finite difference method, and finite volume method. The finite element method is primarily used for structural mechanics and fluid dynamics issues, while the finite difference method is commonly employed for the numerical solution of partial differential equations. On the other hand, the finite volume method gives wide application in fluid dynamics and heat transfer problems. Each method has its applicable scope and advantages, and the choice of an appropriate numerical approach depends on the characteristics of the problem and the requirements of the solution. It is noteworthy that there is a trade-off between the accuracy and computational efficiency of numerical models. Generally, the more refined the model, the greater the computational resources required. Therefore, establishing a suitable numerical model necessitates a balance based on the complexity of the problem and the availability of computational resources.

4.3.1 Construction of numerical model for the LTES heat exchangers

In the construction of numerical models, consideration of parameter selection, grid partitioning, and time step setting is of paramount importance. These choices directly impact the accuracy and the stability of the numerical model. Furthermore, the comparison with actual data and model validation are indispensable steps to ensure the reliability of the numerical model.

4.3.2 Modeling of equations

Numerical models often rely on differential equations, employing numerical methods such as finite element, finite difference, or finite volume techniques to approximate the solutions of these equations. The common numerical model formulas for heat transfer process of the LTES heat exchangers are as follows [10].

Continuity equation:

$$\frac{\partial \rho}{\partial t} + \nabla \cdot \left(\rho \vec{u}\right) = 0 \tag{26}$$

Special attention should be given to grids in regions with density variations, where the density is defined as $\rho = \alpha\rho_l + (1-\alpha)\,\rho_s$. Here, α represents the volume fraction of the liquid phase, defined as:

$$\alpha = \frac{V_L}{V_{cell}} \tag{27}$$

Momentum equation:

$$\frac{\partial \vec{u}}{\partial t} + \nabla \cdot \left(\vec{u}\,\vec{u}\right) = -\frac{1}{\rho}\nabla p + \nu\nabla^2\vec{u} + \beta\rho g\left(T - T_f\right) \tag{28}$$

The parameter T_f represents the reference temperature for buoyancy-driven flow. The second viscosity term is negligible and can be safely disregarded in the model solution [11]. The shear stress can be imposed on the calculation domain through the first term on the right of Eq. (28). The grids in the density-varying region are influenced by the volume fraction α, giving rise to forces arising from density differences with surrounding grids. Consequently, the macroscopic motion of the solid PCM can be directly driven by the stress term.

Energy equation:

$$\frac{\partial h}{\partial t} + \nabla \cdot \left(\vec{u}h\right) = \nabla \cdot \left(\frac{\lambda}{\rho}\nabla T\right) - L\frac{\partial \alpha}{\partial t} \tag{29}$$

Here, h represents enthalpy; λ is the thermal conductivity of the PCM; L is the latent heat of the PCM. The second term on the right side of Eq. (29) accounts for the rate of latent heat change resulting from the solid-liquid phase transition within the grids of the density-varying region.

5. Application cases

5.1 The low-temperature application

The transportation of liquefied natural gas (LNG) is a typical application scenario that can be combined with low-temperature cold thermal energy storage. A large amount of cold thermal energy needs to be released to recover the LNG to the normal atmospheric temperature condition for the end users. Recovering and utilizing this cold thermal energy is of great significance for optimizing the energy consumption of the entire industrial chain. However, it is difficult to achieve a large temperature difference under low-temperature conditions, while maintaining a stable working condition. Therefore, using an LTES heat exchanger to recover and store LNG cold energy becomes rational for real applications. For example, **Figure 7** illustrates the situation of an LNG refrigerated truck, where the LTES heat exchanger serves as a cold storage unit (CSU) integrated with the fuel system of the truck. A large amount of low-temperature cold thermal energy is released during the gasification process and stored for cooling the refrigerated space on the truck. Due to the non-continuous operation of the truck, fluctuations in the cold supply can also be reduced when the LTES heat exchanger is utilized. For the system operation of the refrigerated truck, the LNG is stored in a tank and then released by controlling the valves. It flows into

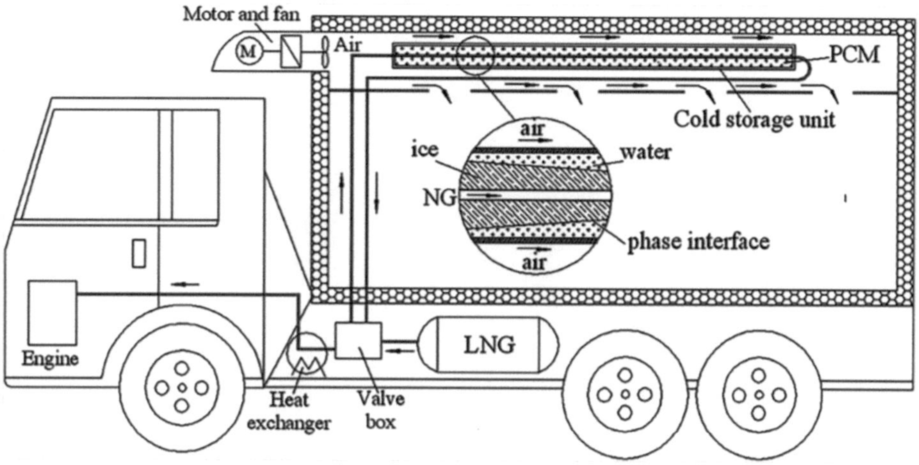

Figure 7.
Schematic diagram of LNG refrigerated truck integrated with the LTES heat exchanger [12].

the CSU to absorb heat from the encapsulated PCM. LNG changes from liquid to gas when passing through the CSU and eventually flows into the vehicle engine for combustion. Due to the low temperature of LNG, which is approximately −162°C, this type of LTES heat exchanger exhibits typical characteristics of large temperature differences and high HTF flow rates. Therefore, the triangular corrugated fins are used to improve the heat transfer process in the application case. The maximum cooling time (or the most effective use of cold thermal energy) can be reached when the aspect ratio of the triangular fins is 0.75. The economical effect can also be realized by using the LTES heat exchanger to store/release cold thermal energy compared to the traditional refrigerated truck.

5.2 The medium-temperature application

A typical scenario for applying the LTES heat exchanger with medium temperature is the heat pump system. The air source heat pump (ASHP) heating technology utilizes the heat energy in the air to provide heating and hot water for end users, which is more energy-efficient and environmentally friendly than traditional heating methods. However, the ASHPs are limited by ambient temperatures during operation, especially in cold regions, where their performance significantly decreases, thus restricting their development. To address the issue of low efficiency of ASHPs in outdoor low-temperature conditions, the LTES heat exchanger combined with solar thermal energy and the ASHP is proposed. The LTES heat exchanger has the advantages of small volume and convenient use, which makes it suitable for applications in small-scale heating scenarios, such as winter heating in rural residences. A typical proposed system can be seen in **Figure 8**. The system consists of solar collectors, hybrid thermal energy storage tanks, circulating pumps, and heat pump systems. When there is sufficient solar radiation outdoors, the thermal energy can be stored in the tank. At night, the thermal energy serves as a heat source for the heat pump, which is conducive to improving the system's energy efficiency and reducing operating costs. The thermal energy storage tank uses RT25 paraffin as the PCM, which has the advantages of high latent heat, large energy storage density, non-toxicity, and non-corrosiveness. The heat exchanger inside the tank adopts a tube-in-tank device

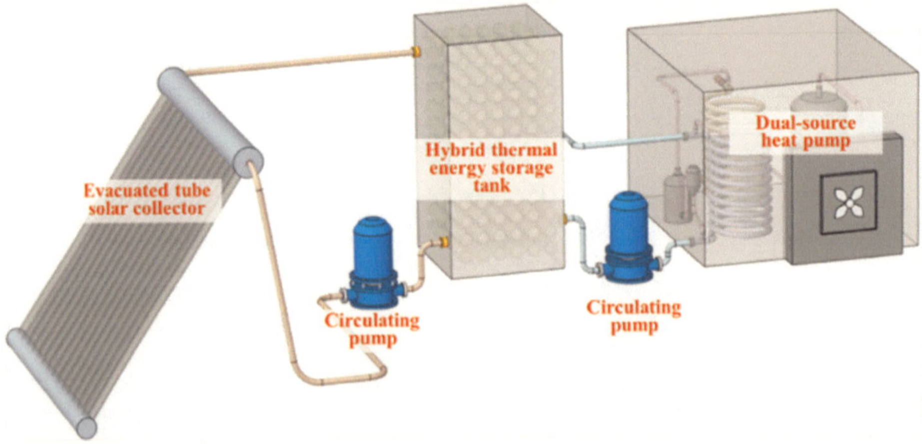

Figure 8.
System composition diagram of the LTES heat exchanger combined with solar thermal collector and ASHP [13].

type. It is estimated that, in northern China, the shortest investment payback period for this system is 5.5 years, which helps to reduce the economic costs of the heat pumps.

5.3 High-temperature applications

In the field of high-temperature thermal energy storage, the thermal storage of power systems has always been a focus of attention. In recent years, electricity production has gradually shifted from conventional thermal power plants to renewable energy sources (RESs). However, the energy generation rate of RESs is closely tied to weather conditions. Due to the high dependence of RESs on nature, the capacity rate is difficult to control. Energy storage systems (ESSs) have attracted attention as they can mitigate the instability of natural resources. In large-scale ESS applications, thermal energy storage systems are favored for their direct utilization of thermal energy. They can store excess heat energy by heating thermal storage mediums. Due to the favorable thermophysical properties such as moderate operating temperatures, high heat capacity, and chemical stability, molten salt has received widespread attention for the development of LTES heat exchangers for power systems. For example, in **Figure 9**, the molten salt TES system is developed to store high-temperature heat to achieve the purpose of power generation.

It is interesting that this device introduces bubble flow to enhance the heat transfer performance of the liquid molten salt during the charging process. The system mainly consists of molten salt tanks, electric heaters, heat transfer pipes, and an air injection system. During the energy storage process, excess electricity from the power grid is converted into heat by multiple electric heaters inside the molten salt tank. During the thermal energy release process, water is flowed into the tank through heat transfer pipes and heated to form the high-pressure steam for electricity generation via steam turbines. To enhance the heat transfer process, multiple serpentine heat transfer pipe bundles are installed inside the storage tank to ensure higher heat transfer rates during the storage and release processes. Additionally, the air injection system serves to further enhance heat exchange by inducing movement of the molten salt around the heat transfer pipes, transitioning the heat transfer mechanism from natural convection to forced convection. This greatly improves the efficiency of the TES system.

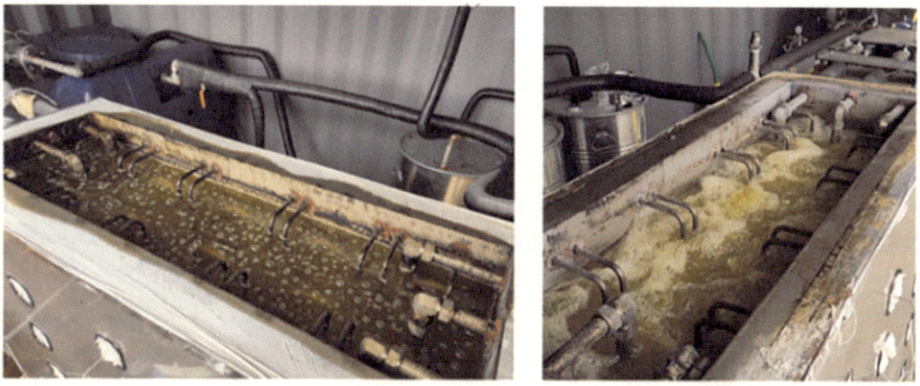

(a) (b)

Figure 9.
Operation of the molten salt energy storage system: (a) static melting; (b) melting with air injection [14].

Therefore, the more efficient high-temperature LTES heat exchanger can effectively alleviate the imbalance between energy supply from the grid and user demand.

6. Conclusions

The LTES heat exchangers provide important advantages in green and sustainable improvement of traditional thermal energy systems by directly storing and releasing thermal energy. The main conclusions of this chapter are summarized below.

In terms of material selections, the basic requirements imposed upon PCMs have been formulated. Due to the diversity of specific applications and the unique characteristics of different types of materials, a large number of materials have been developed. The key properties, including thermal conductivity, PCT, latent heat, and cycling performance, have been extensively tested and validated. However, development and production of commercial materials for LTES heat exchangers are more labor-consuming and long-term processes that demand the attraction of board circles of scientists and experts of various specialties.

In terms of device types, improvements have been made to the HTI, resulting in rich forms of basic structures and improvements. Due to the low heat conductivity of PCMs, surface area density, which is used to present the accessibility of the PCM and the HTF, appears to be an important indicator of structural improvement parameters. In order to improve this parameter, microencapsulation, stable shaping method, and direct contact method are gradually receiving more attention. However, these studies also face important challenges that require further research, mainly focusing on multi-cycle penetration pollution, flow control, and durability of the heat exchanger.

In terms of LTES heat exchanger design, simulation methods from steady-state to dynamic and from one-dimensional to three-dimensional have been developed. There is also mature commercial software available for querying the physical properties of PCMs. However, it still lacks integrated design software for simulation models and physical properties simultaneously, which is mainly required to solve the integration problem of material property queries and also support calculation methods.

In terms of LTES heat exchanger applications, they already have a variety of applications for passive and active heating/cooling scenarios as a component or integrated part of cascaded thermal energy storage systems. More developments and studies about LTES heat exchangers have undoubtedly been directed to low-temperature applications such as district heating/cooling and drying processes. Meanwhile, a detailed investigation is necessary into other applications for high-temperature occasions that can enhance the thermal properties and suppress the supercooling characteristics of PCMs. To sum up, the LTES heat exchangers have been successfully applied in fields such as air conditioning, heat pumps, heating, cold chain, solar power generation, electronic component cooling, and aerospace. It is expected that this type of heat exchanger will gradually attain wider market penetration due to environmental requirements and renewable energy applications.

Acknowledgements

The authors gratefully acknowledge the support from the National Natural Science Foundation of China (No. 52206274).

Conflict of interest

The authors declare no conflict of interest.

Notes/thanks/other declarations

Author contributions: Abstract, Introduction, Materials, and Conclusions, Shen Tian; Device types and variations, Lingling Tian; Design methods, Jiahui Ma; Application cases, Jiawei Li; Funding Acquisition, Shen Tian and Tieying Wang.

Author details

Shen Tian*, Jiahui Ma, Lingling Tian, Jiawei Li and Tieying Wang
Tianjin Key Laboratory of Refrigeration Technology, Tianjin University of Commerce, Tianjin, China

*Address all correspondence to: tianshen@tjcu.edu.cn

References

[1] Nejat P, Jomehzadeh F, Taheri MM, et al. A global review of energy consumption, CO_2 emissions and policy in the residential sector (with an overview of the top ten CO_2 emitting countries). Renewable and Sustainable Energy Reviews. 2015;**43**:843-862. DOI: 10.1016/j.rser.2014.11.066

[2] Hua WS, Lv X, Zhang XL, Ji Z, et al. Research progress of seasonal thermal energy storage technology based on supercooled phase change materials. Journal of Energy Storage. 2023;**67**: 107378. DOI: 10.1016/j.est.2023.107378

[3] Magendran SS, Khan FSA, Mubarak NM, et al. Synthesis of organic phase change materials (PCM) for energy storage applications: A review. Nano-Structures & Nano-Objects. 2019; **20**:100399. DOI: 10.1016/j. nanoso.2019.100399

[4] Palacios A, Navarro-Rivero ME, Zou B, et al. Ding a perspective on phase change material encapsulation: Guidance for encapsulation design methodology from low to high-temperature thermal energy storage applications. Journal of Energy Storage. 2023;**72**:108597. DOI: 10.1016/j.est.2023.108597

[5] Mousa MH, Yang CM, Nawaz K, Miljkovic N. Review of heat transfer enhancement techniques in two-phase flows for highly efficient and sustainable cooling. Renewable and Sustainable Energy Reviews. 2022;**155**:111896. DOI: 10.1016/j.rser.2021.111896

[6] Li NQ, Wang JH, Kleme˅s JJ, et al. A target-evaluation method for heat exchanger network optimisation with heat transfer enhancement. Energy Conversion and Management. 2021;**238**: 114154. DOI: 10.1016/j.enconman. 2021.114154

[7] Yang L, Villalobos U, Akhmetov B, Gil A, Khor JO, Palacios A, et al. A comprehensive review on sub-zero temperature cold thermal energy storage materials, technologies, and applications: State of the art and recent developments. Applied Energy. 2021;**288**:116555. DOI: 10.1016/j.apenergy.2021.116555

[8] Tian S, Yang QF, Hui N, et al. Discharging process and performance of a portable cold thermal energy storage panel driven by embedded heat pipes. Energy. 2020;**205**:117987. DOI: 10.1016/ j.energy.2020.117987

[9] Tian S, Ma JH, Shao SQ, et al. Experimental and analytical study on continuous frozen/melting processes of latent thermal energy storage driven by bubble flow. Energy. 2024;**290**: 130244. DOI: 10.1016/j.energy. 2024.130244

[10] Arıcı M, Tütüncü E, Yıldız Ç, et al. Enhancement of PCM melting rate via internal fin and nanoparticles. International Journal of Heat and Mass Transfer. 2020;**156**:119845. DOI: 10.1016/j.ijheatmasstransfer. 2020.119845

[11] Faden M, König-Haagen A, Höhlein S, et al. An implicit algorithm for melting and settling of phase change material inside macrocapsules. International Journal of Heat and Mass Transfer. 2018;**117**:757-767. DOI: 10.1016/j.ijheatmasstransfer. 2017.10.033

[12] Tan HB, Li YZ, Tuo HF, et al. Experimental study on liquid/solid phase change for cold energy storage of liquefied natural gas (LNG) refrigerated vehicle. Energy. 2010;**35**:1927-1935. DOI: 10.1016/j.energy.2010.01.006

[13] Zhang SL, Liu SL, Wang JH, Li YL, et al. Analysis of a solar-assisted heat pump system with hybrid energy storage for space heating. Applied Thermal Engineering. 2023;**231**:120884. DOI: 10.1016/j.applthermaleng.2023.120884

[14] Che S, Kim J, Jeon J, et al. Thermodynamic analysis of molten salt-based single-tank thermal energy storage system with heat transfer enhanced by gas injection. Journal of Energy Storage. 2024;**77**:109959. DOI: 10.1016/j.est.2023.109959

Chapter 2

Integration of Phase Change Materials in Advancing Heat Exchangers for Enhanced Utilization of Variable Renewable Energy

Aragaw Alamnia

Abstract

Variable renewable energy sources, such as solar and wind power, play a crucial role in sustainable energy systems. However, their intermittent nature poses challenges for maintaining a consistent energy supply. This chapter outlines a research initiative that focuses on the integration of phase change materials (PCMs) within heat exchangers to address these challenges. PCMs, known for their thermal energy storage capabilities, can enhance the efficiency and reliability of renewable energy systems by providing a means to store excess energy during peak generation periods. The chapter aims to explore the synergies between PCMs and heat exchangers, with the goal of optimizing the utilization of variable renewable energy. Through meticulous surveys and integration, this chapter endeavor seeks to overcome the intermittent limitations of renewable energy sources, fostering a more robust and dependable framework for sustainable energy.

Keywords: variable energy, phase change materials, heat exchangers, energy efficiency, sustainable

1. Introduction

Renewable energy sources, such as solar and wind power, play a crucial role in reducing carbon emissions and mitigating the impacts of climate change. Unlike traditional fossil fuels, which can be burned continuously to produce a constant flow of energy, variable renewable energies (VREs) are intermittent, meaning their output fluctuates based on environmental conditions.

Solar power stands as a cornerstone in the transition toward sustainable energy. The harnessing of solar energy involves capturing the sun's rays through photovoltaic cells or solar thermal systems, converting sunlight into electricity or heat. The abundance of sunlight, its widespread availability, and the absence of direct emissions during electricity generation make solar power an environmentally friendly and scalable energy source. However, its intermittent nature, influenced by factors, such as weather conditions and daylight variations, necessitates effective energy storage solutions to ensure a continuous power supply [1].

Wind power represents another critical facet of the renewable energy landscape. By harnessing the kinetic energy of wind through wind turbines, electricity is generated with minimal environmental impact compared to conventional energy sources. Wind power contributes significantly to reducing greenhouse gas emissions and dependency on finite resources. Despite its potential, the variability in wind speed and the unpredictable nature of wind patterns introduce challenges related to energy intermittency [2].

With the global push toward renewable energy sources, the need for efficient energy storage solutions has never been more critical [3]. The intermittent nature of these renewable sources poses significant challenges for grid stability and energy management [4, 5]. For example, during periods of high energy generation, such as sunny days with strong winds, there may be an excess of energy that cannot be effectively stored or utilized. Conversely, during periods of low generation, such as at night or during calm weather, there may be a shortfall of energy, leading to reliance on nonrenewable sources. This imbalance highlights the urgent need for innovative solutions that can store and release energy as needed and integrating phase change materials (PCMs) within heat exchangers offers a promising avenue for addressing this challenge.

The integration of renewable energy sources, into the grid has led to a growing need for effective energy storage solutions [6]. Conventional energy storage methods, such as batteries, have limitations in terms of their capacity, efficiency, and lifespan [7]. For example, lithium-ion batteries, while widely used, are expensive and have limited energy density, making them unsuitable for large-scale energy storage applications. This limitation becomes particularly evident during periods of peak energy generation, where excess energy cannot be efficiently stored for later use. As a result, there is a pressing need for innovative energy storage solutions that can address these challenges and ensure the efficient utilization of renewable energy.

The properties of PCMs to store and release thermal energy this technology offers several key advantages. Firstly, it can significantly improve the efficiency of renewable energy systems by allowing excess energy to be stored and released when needed, thereby reducing waste and increasing overall system efficiency. Secondly, integrating PCMs within heat exchangers can enhance grid stability by providing a reliable and flexible energy storage solution. This is particularly important in regions with high levels of renewable energy penetration, where grid stability can be a major concern. Finally, by enabling more efficient energy storage and utilization, the integration of PCMs within heat exchangers can help reduce carbon emissions and mitigate the impacts of climate change. Overall, this technology has the potential to make a significant impact on the renewable energy landscape, paving the way for a more sustainable and resilient energy future.

The primary objective of this research is to investigate the integration of phase change materials (PCMs) within heat exchangers for the enhanced utilization of variable renewable energy sources. This study aims to explore the feasibility and effectiveness of using PCMs to store and release thermal energy, thereby improving the efficiency and reliability of renewable energy systems. By conducting a comprehensive analysis of the integration process, including the selection of suitable PCMs, design of PCM-integrated heat exchangers, and evaluation of system performance, this research seeks to provide valuable insights into the potential benefits and challenges of this innovative technology. Ultimately, the goal is to contribute to the advancement of renewable energy systems and facilitate the transition toward a more sustainable energy future.

In this chapter, we will explore the integration of phase change materials in advancing heat exchangers for enhanced utilization of variable renewable energy. We will discuss the benefits of this integration, the methods of integrating PCMs into heat exchangers, and the impact on the overall performance of renewable energy systems. Additionally, we will examine case studies and applications where PCM-integrated heat exchangers have been successfully implemented, as well as future trends and research directions in this field.

2. Phase change materials (PCMs)

2.1 Characteristics and properties

Phase change materials (PCMs) are substances known for their ability to store and release thermal energy during phase transitions, such as melting and solidification. This unique property makes them valuable in various applications within renewable energy systems. One key characteristic of PCMs is their high energy storage density, which allows them to store a large amount of energy per unit mass or volume compared to other thermal storage materials.

Additionally, PCMs can regulate temperatures by absorbing or releasing heat at nearly constant temperatures during phase transitions. This feature is crucial for systems requiring stable temperatures, such as in building applications. Moreover, PCMs are thermally stable over many cycles of heating and cooling, making them suitable for long-term use in energy storage applications. Many PCMs are also nontoxic and environmentally friendly, making them suitable for a wide range of applications [8]. Overall, the unique characteristics and properties of PCMs make them invaluable in enhancing the performance and efficiency of renewable energy systems.

The key characteristics and properties of PCMs make them valuable in various applications within renewable energy systems are the following:

- *High latent heat capacity:* PCMs possess a high latent heat capacity, allowing them to absorb or release a significant amount of energy during the phase change process without a substantial change in temperature.

- *Temperature stability:* PCMs maintain a nearly constant temperature during the phase transition, offering a stable thermal environment within a specific temperature range.

- *Reversible phase transitions:* PCMs undergo reversible phase transitions, meaning they can repeatedly change between solid and liquid states without significant degradation, ensuring long-term reliability.

- *Nontoxic and environmentally friendly:* Many PCMs are nontoxic, noncorrosive, and environmentally friendly, making them suitable for various applications, including those in contact with humans and ecosystems.

2.2 Thermal energy storage capabilities

Phase change materials (PCMs) are essential for thermal energy storage due to their unique ability to efficiently capture and release heat energy during phase transitions. When a PCM transitions from a solid to a liquid state (melting) or from a liquid to a solid state (freezing), it absorbs or releases a large amount of energy in the form of latent heat. This property allows PCMs to store thermal energy much more efficiently than traditional sensible heat storage materials, which only store energy through changes in temperature.

One of the key advantages of PCMs is their ability to store energy at a nearly constant temperature during the phase transition. This means that the temperature of the PCM remains relatively stable while absorbing or releasing heat, which is crucial for maintaining a consistent thermal environment. This characteristic makes PCMs particularly suitable for applications where temperature control is critical, such as in buildings, where they can be used to store excess solar heat during the day and release it at night to maintain comfortable indoor temperatures [8–10].

For instance, PCM-enhanced gypsum boards are a practical application of phase change materials (PCMs) in buildings for thermal energy storage. These boards are designed to improve the thermal performance of building envelopes by incorporating microencapsulated PCMs into the gypsum matrix.

During the day, when the outdoor temperature is higher, the PCM absorbs heat and undergoes a phase change from solid to liquid, storing the thermal energy. This process helps to reduce the amount of heat that penetrates into the building, thereby lowering the indoor temperature and reducing the load on air conditioning systems.

At night, when the outdoor temperature drops, the PCM solidifies and releases the stored heat, helping to maintain a comfortable indoor temperature without the need for additional heating. This thermal energy storage and release process can help reduce energy consumption for heating and cooling, making buildings more energy-efficient and cost-effective.

Additionally, PCMs can store a large amount of energy per unit mass or volume, making them space-efficient compared to other thermal storage materials. This high energy storage density allows for compact thermal energy storage systems, which is advantageous in applications where space is limited.

Overall, PCMs are essential for enhancing the efficiency and reliability of renewable energy systems by enabling efficient thermal energy storage. Their ability to store energy with minimal temperature change and high energy storage density makes them invaluable for a wide range of applications from building heating and cooling to industrial processes and energy grid stabilization.

2.3 Types of PCMs

There are various types of PCMs, each with unique properties and applications. Some common types include:

- *Organic PCMs:* Derived from organic compounds such as fatty acids, esters, and alcohols. These PCMs often have a low-melting point and are suitable for medium-temperature applications.

- *Inorganic PCMs:* Composed of salts, metals, or other inorganic materials. Inorganic PCMs typically have higher melting points and are well-suited for high-temperature energy storage applications.

- *Eutectic mixtures:* Combinations of two or more substances that, when mixed in specific proportions, create a mixture with a lower melting point than that of the individual components. Eutectic mixtures are designed for precise temperature control.

- *Bio-based PCMs:* Derived from renewable resources such as plant oils and fatty acids.

Selecting the appropriate type of PCM depends on the specific requirements of the application, including temperature range, thermal conductivity, and cycling stability. The integration of these diverse PCMs into heat exchanger designs holds the potential to revolutionize thermal energy storage in renewable energy systems.

Selecting the right phase change material (PCM) is crucial for integrating PCMs in advancing heat exchangers for enhanced utilization of variable renewable energy (VRE). Several key factors should be considered in the selection process. Firstly, the melting point of the PCM should align with the temperature range of the VRE source and the intended application. PCMs with melting points around 60–70°C are often suitable for solar energy storage. Secondly, the latent heat of fusion, which denotes the energy absorbed or released during the phase change process, is critical. PCMs with higher latent heat values can store more energy per unit volume or mass. Additionally, thermal conductivity is important for efficient heat transfer between the PCM and the heat transfer fluid, maximizing the storage and release of thermal energy. It is also essential to choose a chemically stable PCM that will not degrade or react with other materials in the heat exchanger over its operating temperature range. Compatibility with the encapsulation material, heat transfer fluid, and other components of the heat exchanger is another crucial consideration to prevent issues, such as leakage or degradation. Cost is also a factor, including both the material cost and any additional costs associated with encapsulation or integration. Lastly, the availability of the PCM is important to ensure that it can be readily sourced for maintenance and repair purposes. Considering all those factors, the most suitable PCM can be selected for integration into heat exchangers for enhanced utilization of VRE [8–12].

The three types of phase change materials (PCMs) are suitable for integrating into heat exchangers for enhanced utilization of variable renewable energy (VRE).

Paraffin wax is a commonly used PCM with a melting point range of approximately 45–70°C, making it suitable for storing thermal energy from VRE sources, such as solar or wind power. It has a high latent heat of fusion, allowing it to store a significant amount of thermal energy per unit volume or mass. However, paraffin wax has relatively low thermal conductivity compared to other PCMs, which can affect the heat transfer efficiency in heat exchangers [8, 9]. Nonetheless, its chemical stability and cost-effectiveness make it a practical choice for large-scale applications. Paraffin wax offers a high latent heat of fusion ranging from 150 to 250 J/g, making them ideal for applications requiring high energy storage capacity per unit mass. However, their thermal conductivity is relatively low, typically around 0.2–0.5 W/m K, which can limit heat transfer efficiency in heat exchangers.

For instance, paraffin wax could store 200 J of thermal energy per gram when it undergoes a phase change from solid to liquid. This property makes it well-suited for applications where space is limited but high energy storage is required, such as in portable electronics or building materials [13]. Paraffin wax has a thermal conductivity of 0.3 W/m K, and a melting point of 60°C, which means it is not as efficient at transferring heat compared to other materials suitable for applications where slow and controlled heat transfer is desired in building materials and thermal energy storage systems [14].

On the other hand, salt hydrates, such as sodium sulfate decahydrate or magnesium sulfate heptahydrate, have melting points in the range of 30–100°C, offering flexibility for different VRE applications. They have high latent heat of fusion values and good thermal conductivity, making them effective PCMs for storing and releasing thermal energy. However, salt hydrates can absorb moisture from the air, leading to potential issues with caking and degradation over time. Despite this, their availability and cost-effectiveness make them attractive PCM candidates for medium to large-scale applications [11, 12, 15]. For instance, sodium sulfate decahydrate has a lower latent heat of fusion, ranging from 50 to 150 J/g, but exhibits higher thermal conductivity values, typically between 0.5 and 2.5 W/m K. This makes them more suitable for applications requiring rapid heat transfer. For example, sodium sulfate decahydrate can store 100 J of thermal energy per gram during its phase change, but it can transfer this energy more rapidly than paraffin wax due to its higher thermal conductivity suitable for applications where fast heat transfer is important, such as in heat exchangers or thermal management systems [16]. Sodium sulfate decahydrate has a melting point of 32°C, which means it remains solid at temperatures below this point and transitions to a liquid state above it. It is suitable for applications where higher operating temperatures are required, such as in industrial heat storage or waste heat recovery systems [4, 13, 14, 16].

Bio-based PCMs, derived from renewable sources, offer environmental benefits and sustainability. They have melting points in the range of 5–100°C, providing a wide range of options for different VRE applications [17]. Bio-based PCMs have moderate to high latent heat of fusion values, making them suitable for efficient energy storage [18]. However, they generally have lower thermal conductivity compared to salt hydrates but higher than paraffin wax [19]. Their biodegradability and nontoxic nature make them environmentally friendly options, but they may have lower chemical stability compared to traditional PCMs, requiring proper sealing and encapsulation [20]. For example, eutectic mixtures, such as potassium nitrate-sodium nitrate, offer a unique advantage with a sharp melting point, allowing for precise temperature control. They have a latent heat of fusion ranging from 150 to 300 J/g and moderate thermal conductivity values of around 0.5–1.5 W/m K. Bio-based PCMs, derived from sources, such as palm oil, offer a balance between latent heat of fusion (around 100–200 J/g) and thermal conductivity (approximately 0.1–0.5 W/m K). However, they may be more susceptible to degradation over time compared to synthetic PCMs.

Hence, a eutectic mixture can store 250 J of thermal energy per gram during its phase change, providing a balance between energy storage capacity and temperature control systems [13, 14, 16].

Overall, when selecting a PCM, it should consider the specific requirements of their application. For example, applications requiring high energy storage capacity per unit mass may favor organic PCMs, while those requiring rapid heat transfer may benefit from inorganic PCMs. Eutectic mixtures may be preferred for applications requiring precise temperature control, while bio-based PCMs offer a sustainable and environmentally friendly alternative.

3. Fundamentals of heat exchangers

Heat exchangers are devices designed to transfer heat between two or more fluids, or between a fluid and a solid surface while keeping them physically separated. They play a critical role in various industrial processes and heating, ventilation, and air conditioning (HVAC) systems. The types of heat exchangers and the basic principles of heat transfer involved.

3.1 Types of heat exchangers

- *Shell and tube heat exchangers:* This is one of the most common types of heat exchangers, consisting of a shell (outer vessel) with multiple tubes running through it. One fluid flows through the tubes, while the other fluid flows over the tubes, facilitating heat transfer between the two fluids.

- *Plate heat exchangers:* These heat exchangers consist of multiple thin plates arranged in a way that allows for heat transfer between two fluids. The plates have large surface areas, which enable efficient heat exchange in a compact space.

- *Finned tube heat exchangers:* In this type, extended surfaces (fins) are attached to the outside of tubes to increase the surface area for heat transfer. This design is often used in air-cooled heat exchangers.

- *Double pipe heat exchangers:* This is a simple type of heat exchanger consisting of two concentric pipes through which the hot and cold fluids flow in opposite directions, allowing for heat transfer between them.

Plate heat exchangers and shell and tube heat exchangers are both important for the enhanced utilization of variable renewable energy (VRE) due to their specific characteristics and capabilities, which make them well-suited for this application.

Plate heat exchangers are essential for enhancing the utilization of variable renewable energy (VRE) due to their high efficiency, compactness, and flexibility. Research indicates that plate heat exchangers can achieve high heat transfer coefficients, resulting in efficient heat transfer performance, which is crucial for maximizing the utilization of VRE sources [13]. Additionally, the compact nature of plate heat exchangers makes them suitable for installations where space is limited, such as in renewable energy systems integrated into existing infrastructure [14]. Plate heat exchangers also offer flexibility in design and operation, allowing for easy modification to accommodate changing system requirements [16]. This flexibility is important for adapting to the variable nature of VRE sources.

Shell and tube heat exchangers are also important for the enhanced utilization of VRE, offering versatility, durability, and scalability. Shell and tube heat exchangers are versatile and can handle a wide range of temperatures and pressures, making them suitable for VRE applications where the characteristics of the energy source vary. Additionally, shell and tube heat exchangers are durable and can withstand harsh operating conditions, providing reliable performance over their lifespan. Furthermore, shell and tube heat exchangers can be easily scaled up for larger systems, making them suitable for applications where the energy output of VRE sources is significant [4, 16, 21, 22]. In conclusion, both plate heat exchangers and shell and

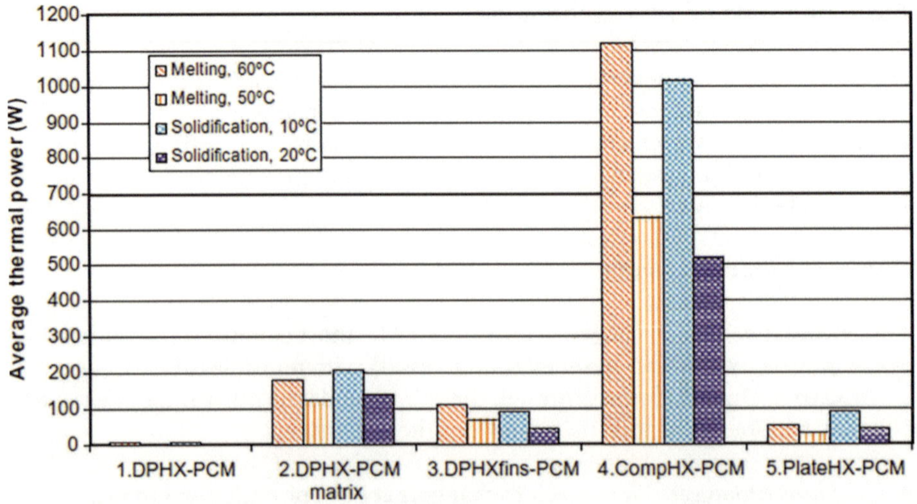

Figure 1.
Normalized thermal power (W/m² K) for charging (melting) and discharging (solidification) processes at two different temperature differences between water and PCM and for the five heat stores studied [23].

tube heat exchangers play a crucial role in maximizing the utilization of VRE and advancing renewable energy systems.

The results in **Figure 1** indicated that fin-tube heat exchangers and double pipe heat exchangers were not suitable for use as heat stores with PCM. The study found that heat exchanger 1 (double pipe heat exchanger with PCM) had a very low heat transfer area, while heat exchanger five (plate heat exchanger with PCM) had a very low ratio of PCM heat capacity over empty heat exchanger heat capacity. On the other hand, heat exchangers two (double pipe heat exchanger with PCM embedded in a graphite matrix), three (double pipe heat exchanger with fins and PCM), and four (compact heat exchanger with PCM) showed better values and were more promising for real applications. Heat exchanger four (compact heat exchanger with PCM) exhibited the highest average thermal power with values above 1 kW for both charging and discharging tests at cases with larger temperature differences between PCM and water. Heat exchanger two (double pipe heat exchanger with PCM embedded in a graphite matrix) showed higher values for average power per unit area and per average temperature gradient, ranging from 700 to 800 W/m^2 K, which were one order of magnitude higher than the ones presented by the second-best heat exchanger. Overall, the study concluded that fin-tube heat exchangers and double pipe heat exchangers were not suitable for integrating PCM in heat storage systems, while other configurations showed more promise for real applications.

Fin-tube heat exchangers (HX) and double pipe heat exchangers are not always the most suitable for integrating phase change materials (PCMs) due to several reasons:

Limited heat transfer area: Fin-tube heat exchangers have a limited heat transfer area per unit volume, which can restrict the amount of PCM that can be effectively integrated, especially if a large amount of thermal energy storage is required. For stance, research by Duffie and Beckman [24] indicates that the heat transfer area of fin-tube heat exchangers is limited by the surface area of the fins, which can restrict the amount of PCM that can be effectively integrated.

Complexity of integration: Integrating PCMs into fin-tube heat exchangers or double pipe heat exchangers can be complex and may require modifications to the existing design, which can increase costs and fabrication time. A study by Cabeza et al. [25] discusses the challenges and complexities involved in integrating PCMs into existing heat exchanger designs, including the need for modifications and potential increases in cost and fabrication time.

Limited thermal conductivity: Fin-tube heat exchangers and double pipe heat exchangers typically have lower thermal conductivity compared to other types of heat exchangers. This can limit the rate at which thermal energy is transferred to or from the PCM, affecting the overall efficiency of the system. Research by Kalogirou [26] highlights that fin-tube heat exchangers and double pipe heat exchangers typically have lower thermal conductivity compared to other types of heat exchangers, which can limit the rate of heat transfer to or from the PCM.

Temperature limitations: Fin-tube heat exchangers and double pipe heat exchangers may not be suitable for applications requiring high-temperature thermal energy storage as the materials used in their construction may not be able to withstand the high temperatures. Studies by Farid et al. [27] and Zalba et al. [28] suggest that the materials used in the construction of fin-tube heat exchangers and double pipe heat exchangers may not be suitable for high-temperature thermal energy storage applications involving PCMs.

Space constraints: Fin-tube heat exchangers and double pipe heat exchangers may have space constraints that limit the amount of PCM that can be integrated, especially in applications where space is limited. Research by Tyagi et al. [29] discusses the space constraints associated with fin-tube heat exchangers and double pipe heat exchangers, which can limit the amount of PCM that can be integrated, especially in applications with limited.

Efficiency concerns: The design of fin-tube heat exchangers and double pipe heat exchangers may not be optimized for efficient heat transfer with PCMs, leading to lower overall system efficiency. A study by Alva et al. [30] investigates the efficiency of heat transfer in fin-tube heat exchangers and double pipe heat exchangers when integrated with PCMs, highlighting potential efficiency concerns compared to other heat exchanger designs.

Overall, those studies provide evidence supporting the notion that while fin-tube heat exchangers and double pipe heat exchangers are commonly used in various applications, they may not always be the most suitable choice for integrating PCMs due to their design limitations and potential efficiency concerns.

4. Basic principles of heat transfer

a. *Conduction:* Heat is transferred through the solid walls of the heat exchanger (e.g., tube walls) by conduction. The rate of heat transfer by conduction is influenced by the thermal conductivity of the material and the thickness of the walls.

b. *Convection:* Heat transfer between a solid surface and a fluid (liquid or gas) flowing over it is primarily by convection. Convection can be natural (due to density differences) or forced (aided by external means such as fans or pumps).

c. *Radiation:* In some cases, heat transfer can also occur through electromagnetic radiation, where heat is emitted or absorbed by surfaces at different temperatures.

The heat transfer equation plays a crucial role in understanding the integration of phase change materials (PCMs) in advancing heat exchangers for enhanced utilization of variable renewable energy (VRE). The basic heat transfer equation governing the heat transfer process is given by:

$$Q = mc\Delta T \tag{1}$$

where Q is the amount of heat transferred, m is the mass of the substance, c is the specific heat capacity of the substance, and ΔT is the temperature change.

For PCMs undergoing phase change, an additional term is included to account for the latent heat of fusion (L):

$$Q = mc\Delta T + mL \tag{2}$$

In the context of heat exchangers, this equation is used to calculate the amount of heat transferred between the PCM and the heat transfer fluid or solid surface. By controlling the flow rate of the heat transfer fluid and the surface area of contact between the PCM and the fluid, the efficiency of heat transfer can be optimized.

The integration of heat exchangers (HX) and phase change materials (PCMs) is governed by several key equations that describe the heat transfer processes within the system. One critical aspect of this integration is the determination of the heat transfer rate between the PCM and the heat exchanger surface, which can be described using Fourier's law of heat conduction.

Furthermore, the Fourier's law of heat conduction is relevant for understanding heat transfer within the PCM:

$$Q = -kAdx \,/\, dT \tag{3}$$

where q is the heat flux, k is the thermal conductivity of the material, A is the cross-sectional area for heat transfer, and $dxdT$ is the temperature gradient.

This equation describes how heat is conducted through the PCM, influencing the rate of temperature change and phase transition within the PCM. By selecting PCMs with high thermal conductivity and designing heat exchangers with optimized surface areas, the efficiency of heat transfer can be improved, enhancing the utilization of VRE in renewable energy systems [13].

The integration of phase change materials (PCMs) in advancing heat exchangers for enhanced utilization of variable renewable energy (VRE) relies on the basic principles of heat transfer. Heat exchangers are devices designed to transfer heat between two or more fluids, or between a fluid and a solid surface to achieve desired thermal conditions. The integration of PCMs enhances this process by leveraging their ability to store and release thermal energy during phase transitions.

One of the fundamental principles of heat transfer in heat exchangers is the concept of thermal conductivity. This property determines the rate at which heat is transferred through a material. PCMs with high thermal conductivity can efficiently transfer heat between the heat source (such as solar or wind energy) and the heat exchanger, ensuring that thermal energy is effectively captured and stored.

Another important principle is heating capacity, which refers to the amount of heat energy required to raise the temperature of a substance by a certain amount. PCMs have a high heat capacity during phase change, allowing them to store large amounts of thermal energy without a significant increase in temperature. This property is crucial for storing excess energy from VRE sources during peak production periods.

Furthermore, the design and configuration of heat exchangers play a crucial role in maximizing the utilization of PCMs. For example, the use of finned heat exchanger surfaces can increase the surface area available for heat transfer, enhancing the efficiency of PCM melting and solidification processes. Additionally, the integration of PCM capsules or containers within the heat exchanger can ensure efficient heat transfer between the PCM and the heat transfer fluid.

In summary, the integration of PCMs in advancing heat exchangers for enhanced utilization of VRE relies on principles of heat transfer, including thermal conductivity, heat capacity, and efficient heat exchanger design. By leveraging these principles, heat exchangers can effectively store and release thermal energy, optimizing the utilization of VRE sources and improving the overall efficiency of renewable energy systems.

The equation for calculating the energy storage capacity is:

$$E = m \cdot L \tag{4}$$

where E is the thermal energy storage capacity, m is the mass of the PCM, and L is the latent heat of fusion of the PCM.

Furthermore, the overall heat transfer coefficient between the PCM and the heat exchanger surface can be determined using the following equation, which considers both the convective heat transfer coefficient (h) and the conductive heat transfer resistance (R):

$$U = 1/(1/h + R) \tag{5}$$

By using these governing equations, can optimize the design of the HX-PCM integration to maximize energy efficiency and enhance the utilization of variable renewable energy sources.

5. Integrating phase change materials

Integrating phase change materials (PCMs) into heat exchangers offers significant benefits, enhancing their efficiency and effectiveness in various applications. One key advantage is the ability of PCMs to store and release large amounts of thermal energy during phase change processes. This capability enables heat exchangers to store excess energy during periods of high availability, such as sunny or windy days, and release it when needed, improving overall energy utilization and grid stability. Additionally, PCMs can help regulate temperatures by absorbing or releasing heat at constant temperatures during phase transitions. This feature can help maintain a more stable temperature profile in heat exchangers, reducing temperature fluctuations in the

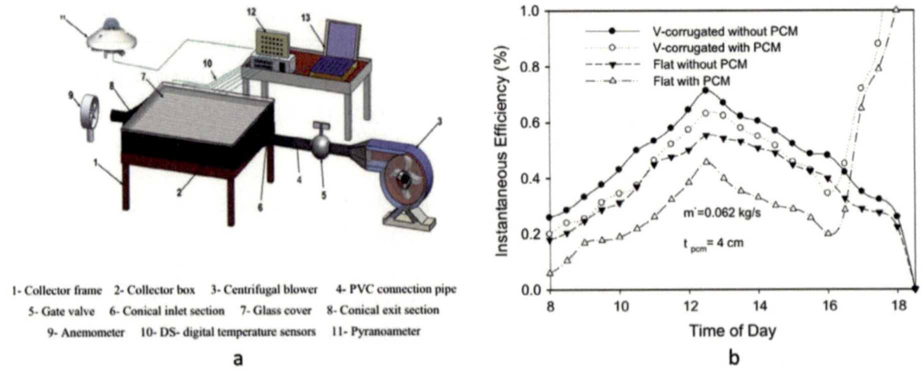

Figure 2.
(a) Schematic diagram of the experimental setup; (b) a comparison of instantaneous thermal efficiency between v-corrugated and flat plate solar air heater with and without PCM when mass flow rate = 0.062 kg/s [17].

system. Moreover, integrating PCMs can improve the efficiency of heat exchangers by reducing the energy required for heating or cooling processes, leading to lower energy consumption and operating costs. Furthermore, PCMs have high energy storage densities compared to conventional thermal storage materials, allowing for more compact designs of heat exchangers and energy storage systems.

As shown above (**Figure 2**), the collector frame provides structural support for the system, holding components, such as the collector box, which houses the heat exchanger and PCM capsules. The centrifugal blower regulates airflow, aiding in heat exchange processes. Polyvinyl chloride (PVC) connection pipes facilitate fluid flow, while the gate valve controls fluid flow rates. The conical inlet and exit sections help direct airflow efficiently. The glass cover protects components and allows for sunlight penetration. Anemometers measure wind speed, pyranometers measure solar radiation, and digital temperature sensors monitor temperatures. Integrating these parts optimizes energy capture and utilization, enhancing the overall performance of renewable energy systems.

The figure comparing the instantaneous thermal efficiency between v-corrugated and flat plate solar air heaters with and without PCM at a mass flow rate of 0.062 kg/s provides insightful comparisons. The inclusion of PCM shows a significant improvement in thermal efficiency compared to systems without PCM. This enhancement can be attributed to the PCM's ability to store and release thermal energy, effectively regulating temperature fluctuations and improving overall energy utilization. Furthermore, the comparison between v-corrugated and flat plate designs indicates that the former generally outperforms the latter in terms of thermal efficiency. This is likely due to the increased surface area for heat exchange in v-corrugated designs, leading to more efficient heat transfer. The selected mass flow rate of 0.062 kg/s allows for a meaningful comparison as it influences the rate of heat transfer and, consequently, the thermal efficiency of the systems. Overall, this figure highlights the importance of PCM integration and collector design in enhancing the efficiency of solar air heaters, providing valuable insights for the optimization of solar thermal systems [4, 14, 16–19, 21, 22].

The integration of phase change materials (PCMs) in advancing heat exchangers for enhanced utilization of variable renewable energy (VRE) involves a carefully designed system setup. A flat plate heat exchanger serves as the core component, featuring inlet and outlet ports for the heat transfer fluid. Inside the heat exchanger, channels facilitate the flow of the fluid with evenly distributed PCM capsules (e.g., paraffin wax) contained within. These capsules are encapsulated to prevent mixing with the fluid. The system includes a control system responsible for regulating the flow rate of the heat transfer fluid and maintaining the temperature within the desired range. Energy from the VRE source (e.g., solar panels, wind turbines) flows to the heat transfer fluid, causing the PCM to melt and store thermal energy. When additional heat is required, the PCM solidifies, releasing the stored energy to the heat transfer fluid. This integration offers benefits, such as improved energy efficiency and grid stability, making it a promising solution for sustainable energy systems [31].

There are several methods to integrate PCMs into heat exchangers, each with its advantages and considerations. One common method is encapsulation, where PCMs are encapsulated within containers or capsules and then integrated into heat exchangers. This approach ensures that the PCM remains contained and does not mix with the fluid in the heat exchanger, preventing contamination. However, proper selection of encapsulation materials is crucial to ensure compatibility with the PCM and heat exchanger operating conditions. Another method is direct contact, where the PCM is in direct contact with the heat transfer fluid (e.g., water, air) in the heat exchanger. As the PCM undergoes phase change, heat is transferred directly to or from the fluid, enabling efficient heat exchange. Proper design is essential to prevent leakage and ensure uniform heat transfer. Additionally, PCMs can be immersed in a heat transfer fluid (e.g., water) within a heat exchanger, allowing for simpler integration compared to encapsulation. However, care must be taken to select PCMs and fluids that are compatible to avoid issues, such as corrosion or degradation. Overall, integrating PCMs into heat exchangers offers a promising pathway to enhance energy efficiency, improve system performance, and promote the integration of renewable energy sources in various applications.

For instance, to integrate paraffin wax into a flat plate heat exchanger for enhanced utilization of variable renewable energy (VRE), several steps must be taken. Firstly, paraffin wax with suitable properties must be selected. For instance, paraffin wax with a melting point between 45°C and 70°C and a high latent heat of fusion is ideal for solar energy storage applications. Next, the flat plate heat exchanger needs to be designed to accommodate the paraffin wax. This involves creating channels for the flow of the heat transfer fluid and ensuring the heat exchanger can withstand the operating temperatures and pressures [15, 17–19]. The paraffin wax should then be encapsulated to prevent it from mixing with the heat transfer fluid. Encapsulation materials should have good thermal conductivity to facilitate heat transfer [32]. Once integrated, a control system is necessary to regulate the flow rate of the heat transfer fluid and maintain the temperature within the desired range for efficient heat exchange with the paraffin wax [33]. In conclusion, integrating paraffin wax into a flat plate heat exchanger requires careful consideration of PCM properties, heat exchanger design, encapsulation, and control systems. Proper integration can significantly enhance the utilization of VRE [34].

6. Enhanced utilization of variable renewable energy

Variable renewable energy sources, such as solar and wind power, face challenges due to their intermittent nature, which can lead to grid instability and difficulty in matching supply with demand. The variability of these sources poses challenges for grid operators in maintaining a reliable electricity supply. However, phase change materials (PCMs) offer a promising solution to mitigate these challenges. They investigated the heat transfer performance of PCMs in a regenerative heat exchanger. Varying Reynolds number, hot fluid temperature, and PCM types revealed increased effectiveness, suggesting potential for designing more efficient and higher-capacity heat exchangers [35].

Researched, Mehling and Cabeza [36] focused on optimizing multiteam plate fin heat exchangers (MPFHE) with four operating streams. Utilizing the non-sorting genetic algorithm (NSGA-II), the study aimed to enhance thermal-hydraulic performance, resulting in up to a 1.5% increase in heat transfer and an 11.1% reduction in power consumption in cryogenic units, promoting efficiency and sustainability. Also, the research explored thermal energy storage using nano-enhanced phase change materials in various plate geometries. Particularly, the sawtooth profile exhibited superior performance, reducing melting time by up to 12% [37].

By integrating PCMs into energy systems, excess thermal energy generated from renewable sources during periods of high availability can be stored. This stored energy can then be released when renewable energy generation is low, helping to balance supply and demand and stabilize the grid.

One of the key benefits of PCMs is their ability to store and release large amounts of thermal energy during phase change processes, such as melting and solidification. Emmi and Bottarelli [19] delved into performance enhancement strategies for heat exchangers, focusing on nanofluids for increased thermal conductivity. Latent thermal energy storage (LTES) heat exchangers, however, lack an analytical method for determining outlet temperature, as addressed by Beyne et al. [17] with an analytical model validated numerically.

This capability allows PCMs to act as a buffer, absorbing excess energy during peak generation periods and releasing it during periods of low generation. This can help reduce strain on the grid and improve overall energy efficiency. Additionally, PCMs can enable load shifting, allowing excess energy generated during off-peak hours to be stored and used during peak demand periods. This can further enhance the utilization of variable renewable energy sources and reduce reliance on fossil fuels. By shifting focus to triple concentric-tube heat exchangers, investigated superior single-phase flow heat transfer characteristics. The study categorized recent publications based on experimental, numerical, and analytical approaches, highlighting advantages, such as shorter length requirements for equivalent performance [38]. Comparing different heat exchanger designs, Kumari and Ghosh [14] evaluated the efficiency of concentric and hairpin heat exchangers for latent heat energy storage using PCMs. Numerical analysis revealed improved heat transfer with reduced diameter and increased length of the high-temperature fluid (HTF).

In phase change heat exchangers, introduced phase-change materials to expand heat dissipation buffer space. Simulations showed that incorporating inner ring ribs into the heat exchange copper tube significantly improved heat transfer efficiency [32].

Overall, integrating PCMs into energy systems offers a promising pathway to enhance the utilization of variable renewable energy sources. By providing a reliable and efficient means of energy storage, PCMs can help address the challenges posed by intermittency and grid instability, paving the way for a more sustainable and resilient energy future.

7. Future trends and research directions

The future of advanced heat exchangers lies in the exploration of emerging technologies and materials that can enhance their performance and efficiency. One promising area is the development of advanced materials with improved thermal conductivity and heat transfer properties. Nanotechnology, for instance, offers the potential to engineer materials at the nanoscale to achieve superior heat transfer characteristics.

Another trend is the integration of smart technologies, such as sensors and control systems, into heat exchangers to optimize their performance and energy efficiency. These technologies enable real-time monitoring and adjustment of heat exchanger operations, leading to more efficient heat transfer processes.

Furthermore, the use of additive manufacturing (3D printing) techniques allows for the rapid prototyping and production of complex heat exchanger designs. This approach opens up new possibilities for designing heat exchangers with optimized geometries for enhanced heat transfer performance.

In terms of research directions, further investigations are needed to explore the potential of phase change materials (PCMs) in improving the energy efficiency of heat exchangers. Studies could focus on developing novel PCM-based heat exchanger designs and materials to maximize heat storage and transfer capabilities.

Additionally, there is a need to explore the integration of renewable energy sources, such as solar and wind power, into heat exchanger systems. Research in this area could lead to the development of hybrid systems that utilize both conventional and renewable energy sources to achieve higher energy efficiency.

Overall, future research in advanced heat exchangers should focus on leveraging emerging technologies and materials to enhance their performance, efficiency, and sustainability. By addressing these challenges, researchers can pave the way for the development of next-generation heat exchangers that play a crucial role in the transition to a more sustainable energy future.

8. Conclusions

In conclusion, this chapter has highlighted the importance of advanced heat exchangers in enhancing the utilization of variable renewable energy sources. The integration of phase change materials (PCMs) into heat exchangers offers significant benefits, including improved heat transfer efficiency and energy storage capacity.

Fundamental principles of heat exchangers, including different types and basic heat transfer mechanisms, were discussed. The chapter also explored various methods of integrating PCMs into heat exchangers, such as encapsulation and direct contact, emphasizing their role in enhancing energy storage and heat transfer processes.

Studies and applications showcased successful integration of PCMs in heat exchangers, leading to performance improvements and energy savings. Numerical analyses and experimental studies provided valuable insights into optimizing heat exchanger designs for improved heat transfer efficiency.

Future trends and research directions were also discussed, highlighting the importance of exploring emerging technologies and materials for advanced heat exchangers. Continued innovation in heat exchanger design is crucial for effectively integrating renewable energy sources into energy systems.

Author details

Aragaw Alamnia
Bahir Dar Institute of Technology, Bahir Dar University, Bahir Dar, Ethiopia

*Address all correspondence to: aragawtadie2007@gmail.com

References

[1] Kenisarin, Mahkamov K. Solar energy storage using phase change materials. Solar Energy. 2007;**81**(1):85-93

[2] Kuznik L et al. Experimental investigation and modeling of a low temperature solar driven heat pump with phase change storage. Energy Conversion and Management. 2011;**52**(1):105-115

[3] Sari S, Kaygusuz N. Thermal conductivity and latent heat thermal energy storage characteristics of paraffin/expanded graphite composite as phase change material. International Journal of Heat and Mass Transfer. 2010;**53**(1-3):170-176

[4] Kakaç S, Liu H. Heat Exchangers: Selection, Rating, and Thermal Design. 2nd ed. CRC Press; 2003

[5] Sharma P et al. Thermal energy storage systems for concentrated solar power plants. Renewable and Sustainable Energy Reviews. 2012;**16**(4):2119-2125

[6] Tatsidjodoung A et al. Thermal conductivity measurement of phase change materials for building applications. Solar Energy Materials and Solar Cells. 2004;**85**(4):333-349

[7] Kabeel AE, Khalil A, Shalaby SM, Zayed ME. Experimental investigation of thermal performance of flat and v-corrugated plate solar air heaters with and without PCM as thermal energy storage. Energy Conversion and Management. 2016;**113**:264

[8] Allahyarzadeh-Bidgoli A, Yanagihara JI. Thermal-hydraulic performance optimization of plate-fin heat exchanger applied in a liquefaction process: Heat transfer and friction investigation. In: International Heat Transfer Conference 17; August 14-18, 2023; Cape Town, South Africa. 2023. DOI: 10.1615/IHTC17.190-40

[9] Garg A, Singhal G. Mathematical modeling for transient thermal performance analysis of phase change material-based heat exchanger. Applied Thermal Engineering. 2022;**216**:119029. DOI: 10.1016/j.applthermaleng.2022.119029

[10] Younis O, Abderrahmane A, Hatami M, Mourad A, Kamel G. Thermal energy storage using nano phase change materials in corrugated plates heat exchangers with different geometries. Journal of Energy Storage. 2022;**55**(Part D):105785. DOI: 10.1016/j.est.2022.105785

[11] Cao Z, Zhang G, Wu Y, Yang J, Sui Y, Zhao X. Energy storage potential analysis of phase change material (PCM) energy storage units based on tunnel lining ground heat exchangers. Applied Thermal Engineering. 2023;**235**:121403. DOI: 10.1016/j.applthermaleng.2023.121403

[12] Güneş T, Şahin M, Kılıç M. Investigation of the effect of different parameters of phase change materials on heat exchanger performance. Heat and Mass Transfer. 2023;**38**(4):1117-1128. DOI: 10.21605/cukurovaumfd.1410784

[13] Das P, Kar SP, Sarangi RK. Review on thermal performance of heat exchanger using phase change material. International Journal of Energy Research. 2022. DOI: 10.1002/er.8345

[14] Kumari P, Ghosh D. A comparative numerical analysis of concentric and hairpin heat exchanger for efficient energy storage using phase-change material. Journal of Thermal Analysis

and Calorimetry. 2023;**148**:12211-12224. DOI: 10.1007/s10973-022-12177-1

[15] El Haj Assad M, Alhuyi Nazari M. Heat exchangers and nanofluids. In: Basu S, editor. Advances in Nanofluids. Elsevier; 2021. pp. 35-60. DOI: 10.1016/B978-0-12-821602-6.00003-1

[16] Zhu C, Lin Z, Liu W, Liu Q, Yan S. Effect of annular ribs in heat exchanger tubes on the performance of phase-change regenerative heat exchangers. Energy Science & Engineering. 2023:1-10. DOI: 10.1002/ese3.1493

[17] Beyne W, Tassenoy R, De Paepe M. Movement of the phase change front in latent thermal energy storage heat exchangers. Journal of Energy Storage. 2022;**57**:106132. DOI: 10.1016/j.est.2022.106132

[18] Akgul D, Mercan H, Acikgoz O, Dalkilic AS. Advancements in triple tube heat exchangers focusing on superior heat transfer characteristics. Kerntechnik. 2023. DOI: 10.1515/kern-2023-0023

[19] Emmi G, Bottarelli M. Improvement of a shallow ground heat exchanger by incorporating phase change materials. Renewable Energy. 2023. DOI: 10.1016/j.renene.2023.02.079

[20] Kannan S, Jog MA, Manglik RM. Experimental study of enhanced heat transfer in phase change material based thermal energy storage in compact heat exchangers. In: ASME 2023 Heat Transfer Summer Conference, V001T01A007. 2023. DOI: 10.1115/HT2023-107111

[21] Manglik RM, Bergles AE. Heat transfer and pressure drop correlations for the rectangular offset strip fin compact heat exchanger. Experimental Thermal and Fluid Science. 2004;**28**(1):77-91

[22] Jacobi AM, Shah RK, Webb RL. Heat transfer augmentation in rectangular channels with vortex generators. In: Proceedings of the ASME-JSME Thermal Engineering Joint Conference. Vol. 3. 1981. pp. 241-248

[23] Medrano M, Yilmaz MO, Nogués M, Martorell I, Roca J, Cabeza LF. Title of the study. Applied Energy. 2009;**86**:2047-2055. Available from: www.elsevier.com/locate/apenergy

[24] Duffie JA, Beckman WA. Solar Engineering of Thermal Processes. 4th ed. Wiley; 2013

[25] Cabeza LF et al. Review on thermal energy storage with phase change: Materials, heat transfer analysis and applications. Applied Thermal Engineering. 2009;**29**(8-9):1809-1838

[26] Kalogirou SA. Solar Energy Engineering: Processes and Systems. Academic Press; 2009

[27] Farid MM, Khudhair AM, Razack SAK. A review on phase change energy storage: Materials and applications. Energy Conversion and Management. 2004;**45**(9-10):1597-1615

[28] Zalba B et al. Review on thermal energy storage with phase change: Materials, heat transfer analysis and applications. Applied Thermal Engineering. 2003;**23**(3):251-283

[29] Tyagi VV et al. Review on thermal energy storage with phase change: Materials, heat transfer analysis and applications. Renewable and Sustainable Energy Reviews. 2015;**14**(1):167-188

[30] Alva G et al. Thermal Energy Storage Systems and Applications. John Wiley & Sons; 2018

[31] Shah RK, London AL. Laminar Flow Forced Convection in Ducts. Academic Press; 1978

[32] Zhou Z, Zhang Z, Zuo J, Huang K, Zhang L. Phase change materials for solar thermal energy storage in residential buildings in cold climate. Renewable and Sustainable Energy Reviews. 2015;**48**:692-703

[33] Ma T, Yang H, Zhang Y, Lu L, Wang X. Using phase change materials in photovoltaic systems for thermal regulation and electrical efficiency improvement: A review and outlook. Renewable and Sustainable Energy Reviews. 2015;**43**:1273-1284

[34] Seddegh S, Wang X, Henderson AD, Xing Z. Solar domestic hot water systems using latent heat energy storage medium: A review. Renewable and Sustainable Energy Reviews. 2015;**49**:517-533

[35] Morrison D, Abdel-Khalik S. Effect of phase-change energy storage on the performance of air-based and liquid-based solar heating systems. Solar Energy. 1978;**20**:57-67

[36] Mehling H, Cabeza LF. Heat and Cold Storage with PCM. New York: Springer; 2008

[37] Kamali S. Review of free cooling system using phase change material for building. Energy and Buildings. 2014;**80**:131-136

[38] Jankowski NR, McCluskey FP. A review of phase change materials for vehicle component thermal buffering. Applied Energy. 2014;**113**:1525-1561

Section 2

Modeling Methodologies

Chapter 3

Recent Progress in Heat Transfer and Fluid Flow Characteristics within Cross-Flow Heat Exchangers: A Passive Approach

Arash Mahdavi

Abstract

Efficient heat transfer in industry and commercial sectors is vital if saving energy is a concern. Heat exchangers are devices that are built for this aim. Here, cross-flow heat exchangers are analyzed to increase the heat transfer rate and reduce the pressure drop simultaneously. Cross-flow heat exchangers possess a simple geometry, yet a complex flow behavior. The reason lies behind the von Karman vortex structures that are created due to the presence of an initial row of tubes. This will affect the next stages of tube bundles and form an intricate thermo-hydraulic interplay. In general, there are various passive techniques to enhance the efficiency of cross-flow heat exchangers such as tube orientation, adding vortex generators, and addition of fins. Each technique has its strengths and weaknesses that try to address them individually in this chapter.

Keywords: heat exchanger, passive techniques, heat transfer, thermo-hydraulic performance, fluid dynamics

1. Introduction

Heat exchangers play a crucial role in various industries and commercial sectors by efficiently transferring heat between two or more fluids at different temperatures while keeping them physically separate [1, 2]. Their applications range from heating, ventilation, and air conditioning (HVAC) systems in buildings to large-scale industrial processes like power generation, chemical manufacturing, petroleum refining, and food processing [3, 4]. One of their primary functions is to either cool or heat fluids, depending on the process requirements, to maintain optimal operating conditions and enhance energy efficiency.

There are several types of heat exchangers commonly used across industries, each with its advantages and limitations. One widely used type is the shell and tube heat exchanger, which consists of a bundle of tubes enclosed within a cylindrical shell [5].

This design allows for high heat transfer efficiency and versatility in handling a wide range of fluids and operating pressures. Another common type is the plate heat exchanger, which utilizes a series of thin metal plates to facilitate heat transfer between fluids. Plate heat exchangers with high performance-to-volume ratios are compact, lightweight, and offer excellent thermal performance, making them suitable for applications where space and weight are limited [6].

Cross-flow heat exchangers are a specific type of heat exchanger where fluids flow in orthogonal directions. They consist of fluid flow across a bundle of tubes, leading to the formation of boundary layers, flow separation, and vortices.

These heat exchangers are widely adopted in the industry due to their simple and compact design, efficient heat transfer, reduced fouling, ease of maintenance, versatility across a wide range of operating conditions, cost-effectiveness, lower pressure drop, and flexibility in fluid selection [7]. Although cross-flow heat exchangers possess a seemingly uncomplicated architecture, their fluid dynamics exhibit intricate complexities owing to Von Kármán-like oscillations occurring in the wake of tubes and subsequent collisions. Notably, the flow and thermal characteristics within the initial row closely resemble those surrounding a solitary cylinder in cross-flow conditions [8]. Nevertheless, as the flow progresses through successive rows, instabilities manifest and propagate downstream. These instabilities precipitate heightened turbulence levels, consequently augmenting heat transfer rates within the inner rows [9]. Improving the efficiency of heat exchangers stands out as the paramount priority for maximizing their operational effectiveness across all varieties. A range of methods, both active and passive, can be employed to enhance heat transfer efficiency [10]. Numerous research studies have been conducted on improving the performance of cross-flow heat exchangers using passive techniques that require no extra pumping power compared to active methods [11]. These techniques include the use of vortex generators, winglets, or the addition of fins [12]. Additionally, implementing oval-shaped [13] or cam-shaped tubes [14], as shown in **Figure 1**, is another option for reducing the pressure drop. Since increasing the heat transfer rate is often accompanied by an increase in pressure drop, many researchers have investigated and optimized heat exchanger designs to achieve optimum performance.

Numerous review articles are published, aiming to consolidate passive techniques. When aiming for optimal heat exchanger selection, key considerations include design and operational parameters. This review exclusively delves into passive heat transfer

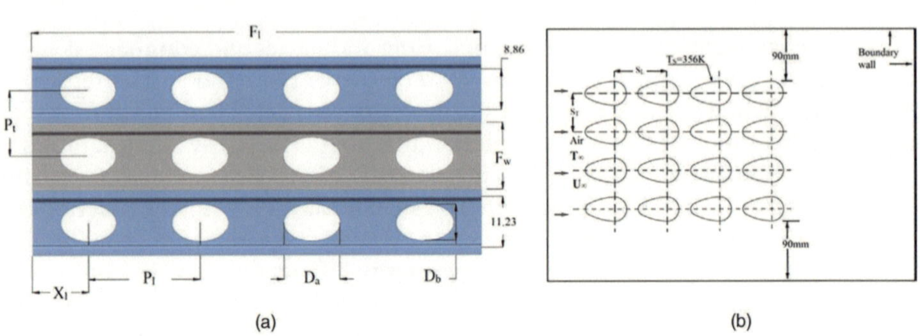

Figure 1.
Pictures of oval-shaped tubes [15] and cam-shaped tubes [16].

techniques, addressing significant areas of interest and performance aspects. The following section discusses various types of heat transfer enhancement methods for cross-flow heat exchangers.

2. Passive heat transfer enhancement methods

Passive techniques implemented to enhance the heat transfer rate of a cross-flow heat exchanger include tube geometry orientation, vortex generators, and the addition of fins to the tube bundle. In this section, we will review the relevant research conducted in each area of passive cooling methods applied to cross-flow heat exchangers. Although many of the techniques mentioned earlier offer the common advantage of enhancing heat transfer rates, their drawbacks include increased complexity and manufacturing costs, as well as elevated pressure drops.

2.1 Tube geometry orientation

In the realm of thermal dynamics, there has been a concerted focus on innovating tube cross-sections to optimize heat transfer efficiency. While circular shapes have long been the norm, researchers have ventured into uncharted territories, exploring streamlined tube designs for improved energy exchange. Shapes ranging from elliptical to tubular configurations have been under scrutiny, each promising potential advancement in cross-flow scenarios.

A seminal study by Buyruk [17] delved into the intricacies of heat transfer dynamics amidst cross-flow scenarios, employing advanced numerical simulations through finite element methodologies. The experimentation involved manipulating tube pitch ratios across a spectrum, from 1.13 to 6, all while holding the Reynolds number constant at 400. The computational framework of this investigation is illustrated in **Figure 2**, shedding light on the intricacies of energy interaction in this domain.

Beyond the confines of traditional circular cross-sections, this exploration aims to revolutionize thermal engineering practices, offering a pathway toward

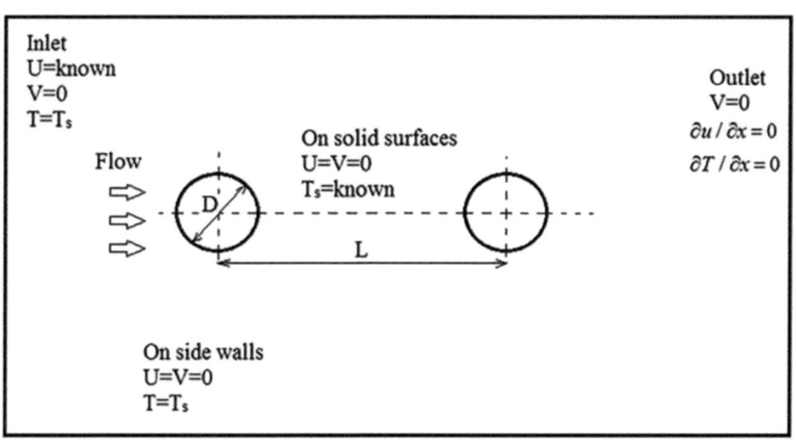

Figure 2.
Computation domain of reference [17] study.

unprecedented efficiency gains. Through meticulous analysis and experimentation, researchers endeavor to unlock the full potential of alternate tube geometries, ushering in a new era of thermal optimization.

Merker et al. [18] undertook an experimental exploration concerning heat transfer and pressure drop in an offset tube array employing elliptical tubes. Their approach involved employing the mass transfer analogy, utilizing naphthalene as the medium for analysis. Oval tubes exhibited relatively lower pumping power requirements compared to traditional circular tubes, owing to their reduced frontal area.

In a parallel study, Aiba [19] delved into the heat transfer dynamics of straight round tubes positioned near a plane wall, covering a broad spectrum of Reynolds numbers from 8000 to 40,000. Nusselt numbers were derived by manipulating variables such as clearance between the walls and the ratio of tube spacing to tube diameter.

Likewise, Toolthaisong and colleagues [20] investigated the impact of attack angles on the thermal performance and pressure differential on the air side in a cross-flow heat exchange setup featuring an offset arrangement of flat tubes, illustrated in **Figure 3**.

Circular tubes exhibit superior heat transfer rates compared to flat tubes, whereas flat tubes are favored for their ability to minimize pressure drop. This necessitates an investigation into the optimal aspect ratio and tube pitch ratio for flat tubes to optimize overall thermal performance. This examination should consider the integrated effects of both heat transfer and pressure drop amidst the tube clusters. In this context, Horvat et al. [21] examined different geometries of tube cross-sectional shapes to investigate their impact on heat transfer rates. Utilizing the documented temporal patterns of speed and heat, the average temporal values of the Reynolds figure, resistance factor, and Stanton factor were computed. Their findings indicated reduced resistance factor and Stanton factor for ellipsoid tubes in contrast to cylindrical counterparts. Nonetheless, with the augmentation of the hydraulic dimension, the distinction diminishes between the dual tube configurations. The universal pattern for resistance factor and Stanton quotient remains steady throughout all structures. Nevertheless, both the resistance factor and Stanton quotient dwindle with the surge in Reynolds magnitude. The results suggest that ellipsoidal and wing-shaped tubes exhibit decreased resistance factor and Stanton quotient in comparison to cylindrical tubes.

In an exhaustive study by Berbish [22], the intricate interplay of heat transfer and flow dynamics surrounding four staggered elliptic cylinders in crossflow was meticulously examined, as shown in **Figure 4**. These cylinders, characterized by an aspect

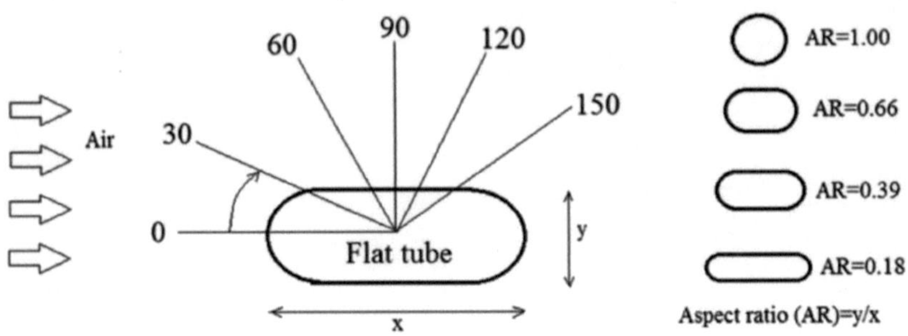

Figure 3.
Various attack angles and aspect ratios studied in Ref. [20].

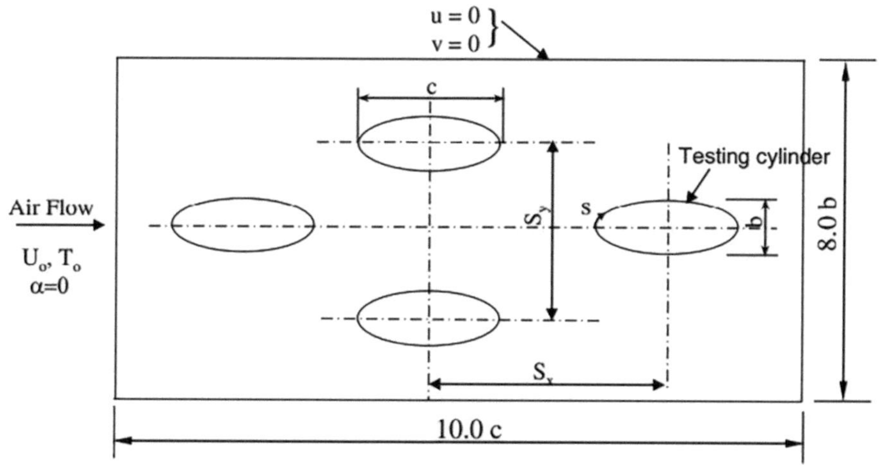

Figure 4.
Computational domain and arrangements of elliptic tubes in Ref. [22].

ratio of 1:2, were meticulously arranged at zero attack angle relative to the incoming flow. Heat transfer data were meticulously gathered solely from the downstream elliptical cylinder, termed the test cylinder, under meticulously controlled heat flux conditions, with air serving as the primary test medium. An expansive range of Reynolds numbers, spanning from 4000 to 45,570, predicated on the major axis length (c), was meticulously scrutinized, alongside longitudinal spacing ratio (S_x/c) and lateral spacing ratio (S_y/b) meticulously varied between 1.5 and 4.0. Employing state-of-the-art computational fluid dynamics (CFD) software, intricate airflow patterns and temperature fields enveloping the cylinders were meticulously simulated, while meticulous flow visualization techniques were meticulously employed to vividly illustrate the complex flow features. Surpassing mere observation, the study unveiled remarkable insights, notably revealing that, barring Reynolds number 4000, the Nusselt number (Nu) of the downstream elliptic cylinder in the four staggered arrangement conspicuously surpassed that of the three inline cylinders across all meticulously tested spacing ratios and Reynolds numbers. Intriguingly, at lower Reynolds numbers ($Re < 14{,}100$), the average Nusselt number of the downstream elliptic cylinder in the three staggered arrangement markedly exceeded that of the downstream cylinder in the four staggered layout, across all meticulously tested spacing ratios. Yet, as the Reynolds numbers scaled to greater than or equal to 14,100, a conspicuous shift occurred, with the staggered layout boasting discernibly elevated mean Nusselt number values for the scrutinized elliptical cylinder. This meticulous delineation underscores the profound influence wielded by cylinder arrangement and Reynolds numbers on the intricate tapestry of heat transfer characteristics in such convoluted configurations.

2.2 Vortex generators

Winglets and vortex generators are designed to manipulate the flow field around surfaces, mitigating flow separation. By controlling the formation and behavior of vortices, they help maintain attached flow, which improves overall efficiency and performance. In heat exchangers, winglets or vortex generators can enhance heat transfer rates by promoting mixing and increasing convective heat transfer coefficients. They disrupt the boundary layer, enhancing the exchange of heat between the fluid and the solid surface.

Xu et al. [23] devised an elliptical fin-and-tube heat exchanger incorporating X-VGs (cross-shaped vortex generators) to enhance both hydraulic and heat performance, as shown in **Figure 5**. Their findings indicate that increasing the streamwise distance and the main flow angle leads to a reduction in thermal-hydraulic performance. However, there exists a non-linear correlation between longitudinal placement and thermal-hydraulic performance.

To determine the optimal layout of X-VGs, the researchers employed a Genetic Algorithm combined with a radial basis function neural network prediction model. The best layout was achieved with a streamwise space ratio of 1.995, longitudinal placement ratio of 0.387, and main attack angle ratio of 0.053. This layout yielded a remarkable 109.681% enhancement in performance evaluation criteria compared to the baseline case.

Torii et al. [24] undertook empirical investigations to analyze the heat transfer and pressure differential attributes linked with winglet-style vortex generators integrated

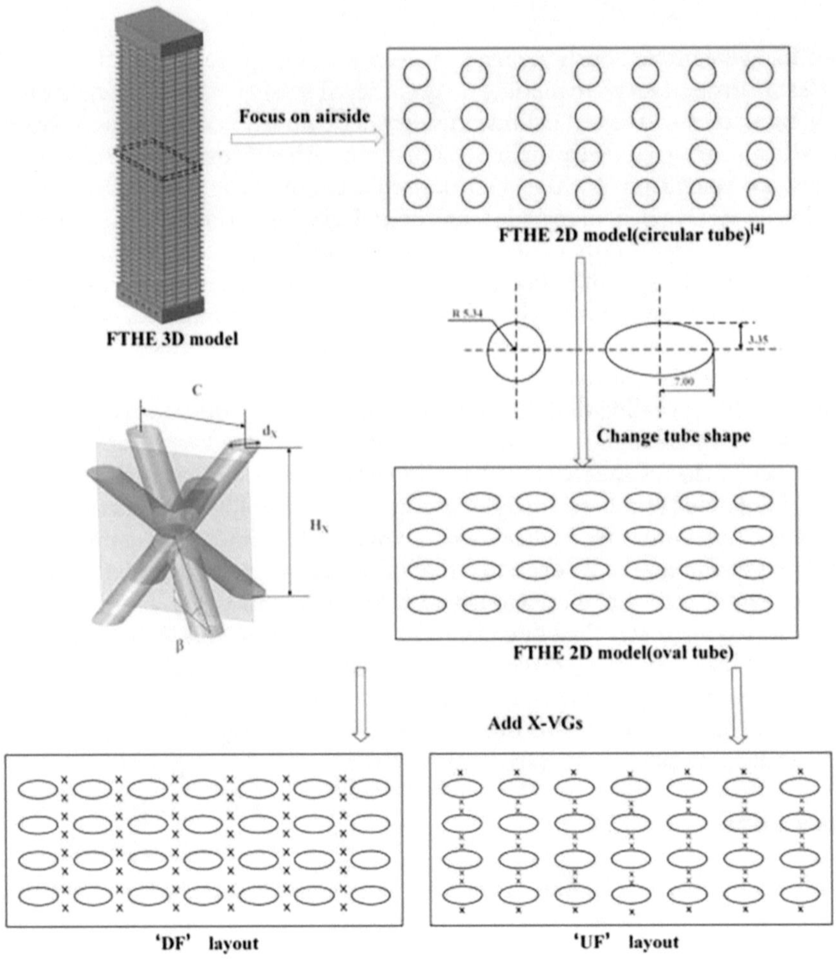

Figure 5.
Geometric depiction of 3d and 2d views of reference [23].

into fin-tube heat exchangers. Both conventional upward flow and downward flow configurations were explored. **Figure 6** illustrates a schematic depiction of the vortex generators. The Reynolds number was consistently kept within the range of 250–2100 throughout the experiment. Comparing the in-line and staggered configurations of the tube bank, there was observed a notable enhancement in heat transfer, approximately ranging from 10 to 20% and 10 to 30%, respectively. Moreover, an increase in pressure drop was also noted, with percentages varying from 8 to 15% for the inline configuration and from 34 to 55% for the staggered configuration.

Saini et al. [25] conducted a numerical investigation on a three-dimensional fin-and-tube heat exchanger, incorporating curved delta winglets as vortex generators (**Figure 7**). The Reynolds number ranged between 400 and 2000. The study analyzed the pressure distribution, temperature distribution, and flow structure distribution of the heat exchanger with a four-in-line circular tube configuration, comparing it with a heat exchanger without vortex generators.

The heat exchanger equipped with vorticity inducers experiences a significant increase in the Nusselt number, with enhancements of 77.25 and 42.51% observed at

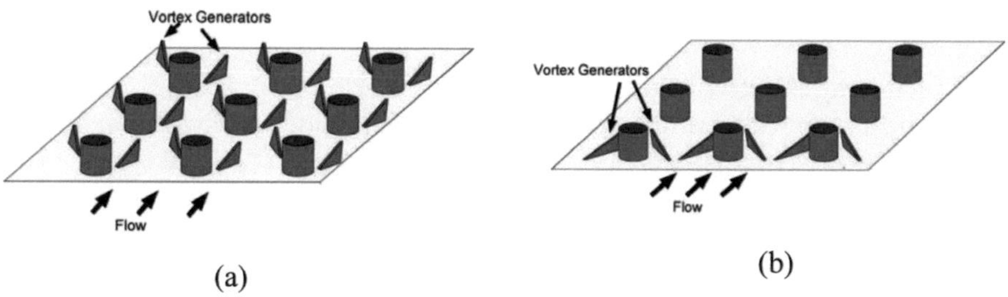

Figure 6.
Configuration of winglet-type vortex generator on the fin surface-tube bank [24].

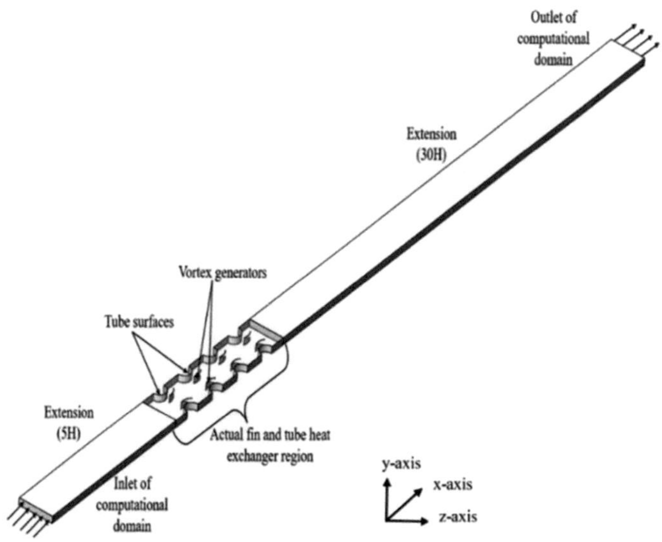

Figure 7.
Three-dimensional computational domain of fin-and-tube heat exchanger as described in Ref. [25].

Reynolds numbers of 400 and 2000, respectively, compared to a fin-and-tube heat exchanger without vortex generators. Additionally, friction is reduced by 5.11%.

In another study focusing on winglet geometries, Zhao et al. [12] investigated nine different cases involving curved and straight rectangular winglet vortex generators integrated into a fin-and-tube cross-flow heat exchanger. When comparing the cases with winglet vortex generators to those without, the Colburn factor was consistently higher across all case studies. Furthermore, the findings indicated that curved winglets with smaller offsets exhibited better overall performance (**Figure 8**).

While numerical calculations are more cost-effective than expensive experimental tests, achieving the optimal performance of a cross-flow heat exchanger at a specific design configuration entails a substantial computational cost. Hence, Xie et al. [26] employed response surface methodology in conjunction with artificial neural networks to determine the optimal design of a cross-flow heat exchanger equipped with vortex generators. According to their findings, both optimization techniques concluded that minimizing the attack angle to 160 degrees and maximizing the arc angle to 10 degrees would result in optimal performance, taking into account both heat transfer and friction factor simultaneously.

2.3 Addition of fins

Enhancing the effective heat transfer area is a traditional approach that leads to increased heat transfer rates. The widespread use of extended surfaces is well-known across industries, particularly in cross-flow heat exchangers. Different types of fins, including plate fins, rectangular straight fins, helical fins, and others, are commonly employed in cross-flow heat exchangers. In this review, we will endeavor to examine research conducted on cross-flow heat exchangers equipped with extended surfaces or fins.

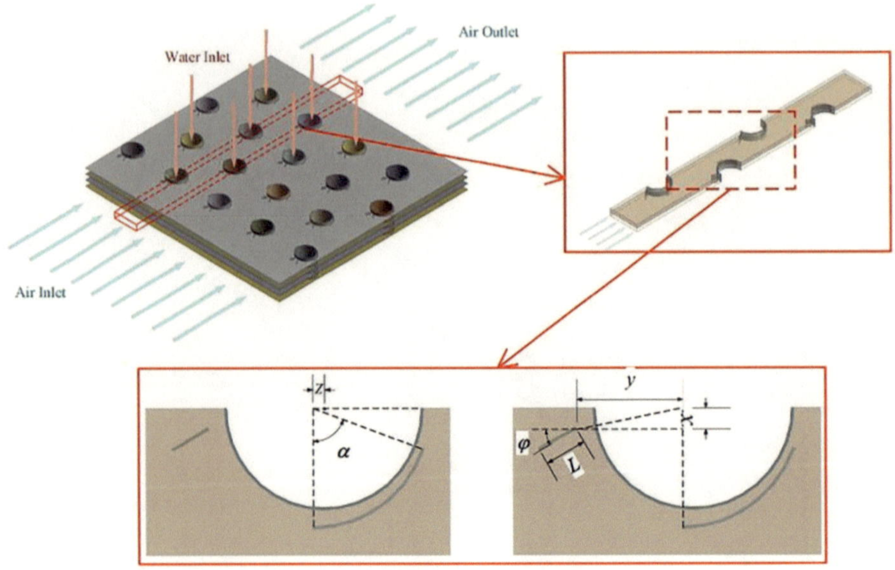

Figure 8.
Schematic view of a heat exchanger with rectangular winglet vortex generator described in Ref. [12].

Mitra et al. [27] conducted a study on the conjugate heat transfer of a finned tube, as depicted in **Figure 9**, specifically focusing on circular tube configurations. Their analysis encompassed the evaluation of flow patterns, pressure distributions, Nusselt numbers, and fin efficiencies. This setup reflects the conventional alignment of fins on the tube surface, depicting the directions of hot and cold fluids, which mirrors the simplest form of fin tube configuration.

Sparrow and Kang [28] presented a simplified approach, suggesting the placement of a single fin at the front, back, or both ends for circular tubes. The diagram illustrating the arrangement of the fins is presented in **Figure 10**. The Reynolds number for the flow across the fluid ranged up to 8000. The results suggested that adding longitudinal fins to the tube improves the heat transfer rate while simultaneously decreasing pressure drop, especially for fins affixed to the rear of the cylinder in contrast to tubes without fins, conversely when fins are attached to both the front and rear ends, both the heat transfer rate and pressure drop rise.

Martinez et al. [29] analyzed four semi-empirical models of heat transfer and pressure drop for helically segmented finned tubes in a staggered layout. Their study focused on assessing the performance of a helically segmented finned tubes heat exchanger on an industrial scale and comparing predictions with experimental

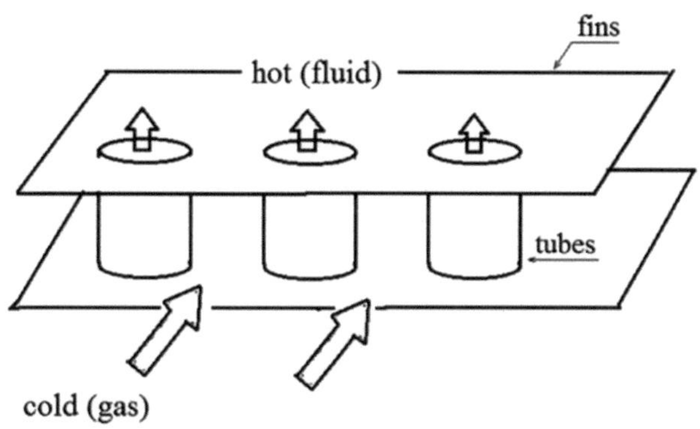

Figure 9.
Cross-flow heat exchanger equipped with plate fins [27].

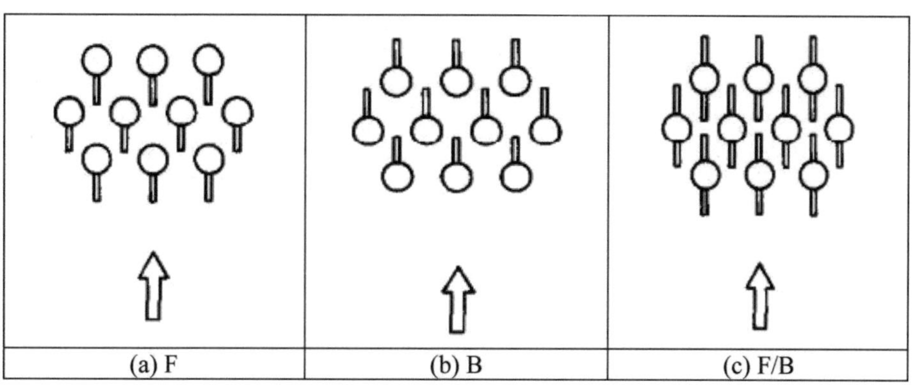

Figure 10.
Finned tube heat exchanger used in work [28] with different configurations of (a) front, (b) back, and (c) front and back.

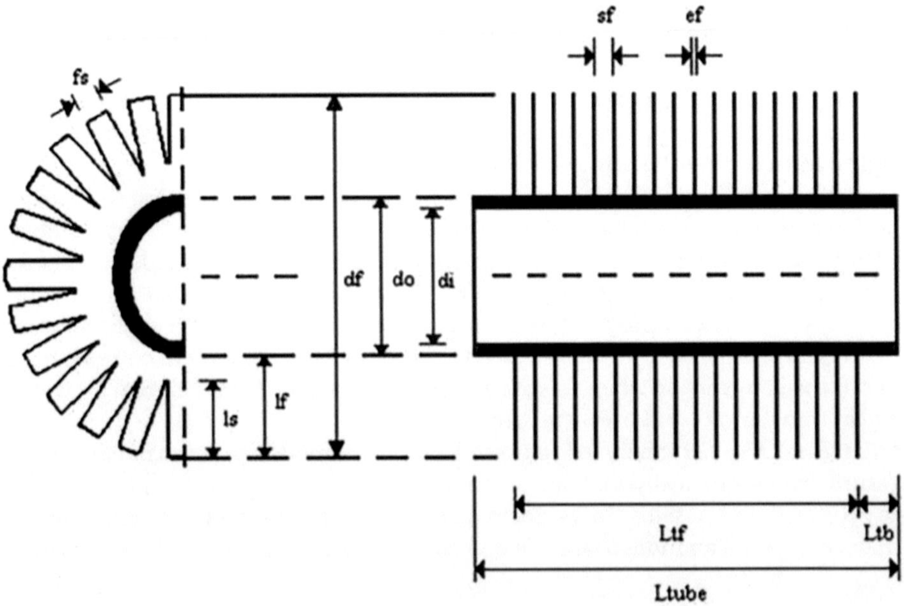

Figure 11.
The schematic representation of the helically segmented tube by Martinez et al. [29].

data. Utilizing the logarithmic mean temperature difference (LMTD) method for thermal analysis, they found that a combination of the Kawaguchi and Gnielinski models achieved a precision greater than 95% in heat transfer prediction at a flue gas Reynolds number of approximately 10,000. Regarding pressure drop, the Weierman model exhibited a precision of around 90% at the same Reynolds number. Overall, their results indicated that the optimal flow regime, achieving optimal heat transfer and pressure drop, was observed at a Reynolds number of about 10,000 based on the outside bare tube. Their schematic view of the helical segmented tube is shown in **Figure 11**.

3. Lessons summary

Given the amount of research done on ways to improve heat transfer passively in cross-flow heat exchangers, it is important to summarize all the articles mentioned in Section 2. **Table 1** summarizes the papers, noting their main findings, their study type, and the parameters they tested. Plus, since we mention lots of case studies in the next table, it might be helpful to quickly go over the design details and what they achieved, as well as how we could combine different passive methods for future work. This next table aims to make it easier to compare and classify different ways to cool passively.

4. Conclusion

Following an in-depth analysis of recent developments in cross-flow heat exchangers, numerous conclusions and recommendations emerge:

Researcher	Type of study	Tested parameters	Results
Sparrow and Kang [28]	Experimental (addition of longitudinal fins)	Placement of the fins, fin tip shape, and the fin thickness	Compared to greater diameter tubes, finned tubes result in a higher heat transfer rate at a given pressure drop
Merker and Hanke [18]	Experimental (tube geometry orientation)	Varying transversal and longitudinal pitches of staggered tubes	Oval-shaped tube banks are desired compared to circular tube banks
Aiba [19]	Experimental (tube geometry orientation)	Cylinder diameter, clearance between the cylinders, and pitch between cylinders	A correlation for overall heat transfer was yielded based on the given parameters
Chen et al. [27]	Numerical (tube geometry orientation and addition of fins)	Reynolds number and fin parameter	A horseshoe vortex was observed near the oval-shaped tube, and the formed drag due to the tube shape was small
Buyruk [17]	Numerical (tube geometry orientation)	Longitudinal distance of cylinders and Reynolds number	Flow acceleration resulted in a higher local heat transfer coefficient for the second cylinder compared to the initial row
Torii et al. [24]	Experimental (addition of fins and vortex generators)	Reynolds number	Heat transfer was enhanced by 10–30% while pressure loss was reduced by 34–55%
Horvat and Mavko [21]	Numerical (tube geometry orientation)	Varying diagonal pitch-to-diameter ratio	The drag coefficient and Stanton number decrease with an increase in Reynolds number
Martinez et al. [29]	Experimental (addition of fins)	Reynolds number	Weierman model came up with the best accuracy for Reynolds number above 10,000
Berbish [22]	Experimental and numerical (tube geometry orientation)	Longitudinal spacing ratio, transversal spacing ratio, and Reynolds number	The average Nusselt number is greater for the single elliptic cylinder compared to the circular one
Toolthaisong and Kasayapanand [20]	Experimental (tube geometry orientation)	Air attack angles, tube aspect ratios, and Reynolds number	By increasing the attack angle, both heat transfer and pressure drop were increased
Xie et al. [26]	Numerical (addition of fins and vortex generators)	Length, arc angle, and attack angle of vortex generator	At an attack angle of 160 degrees and arc angle of 10 degrees, the heat transfer was maximized
Saini et al. [25]	Numerical (addition of fins and vortex generators)	Number of punch holes in vortex generators and Reynolds number	The curved delta winglet vortex generator addition dwindled the size of the wake downstream
Zhao et al. [12]	Numerical (addition of fins and vortex generators)	Arc angle, attack angle, offset, and Reynolds number	The heat exchanger performance was improved by combining straight and curved winglets
Xu et al. [23]	Numerical (addition of fins and vortex generators)	Main angle, streamwise distance, longitudinal placement, and Reynolds number	A 10-degree main angle showed the optimum performance based on the radial basis function neural network

Table 1.
The passive methods applied to cross-flow heat exchanger.

1. Tube geometry orientation:

    - Employing non-circular tubes such as elliptical, wing, or cam-shaped configurations, particularly when arranged in an offset pattern with a decreased angle of fluid flow, enhances heat transfer efficiency.

    - Elliptic or oval tubes, available commercially with varying aspect ratios, offer a promising alternative to circular tubes, providing superior thermal performance and lower air side pressure drop.

    - While more advanced tube geometries like cam-shaped and lenticular tubes offer enhanced streamlined surfaces, their complex manufacturing processes may limit widespread industrial application.

    - Optimal thermal performance, balancing heat transfer and friction factor, is achieved at lower attack angles. Excessive streamlined tube designs should be avoided to ensure sufficient turbulence generation.

2. Vortex generators:

    - Vortex generators, including delta and rectangular winglets, significantly augment heat transfer by inducing additional turbulence in the cross-fluid path.

    - Delta winglet vortex generators are particularly effective, generating secondary flows alongside the mainstream flow. They can be easily mounted on fin surfaces, simplifying manufacturing processes.

    - Curved triangular and box-shaped winglet vorticity inducers offer the additional advantage of directing fluid movement over tubes, simultaneously diminishing wake dimensions and creating secondary flow.

    - Control of higher pressure drops can be achieved by implementing winglets on alternate rows, especially in staggered tube configurations.

3. Addition of fins:

    - Utilizing fin surfaces enhances heat transfer by expanding the surface area available for heat exchange. Helical, plate and longitudinal fins represent commonly utilized options, with continuous plate fins being favored for larger tube banks owing to their lower pressure drop.

    - Thin plate continuous fins are favored by researchers for their simplicity and ease of mounting, allowing for additional passive heat transfer methods if needed. Conversely, helical fins may present constraints due to tube pitch ratios.

    - Thinner fin thicknesses offer advantages, as both fins and tubes maintain similar temperatures, thereby enhancing heat transfer rates while reducing pressure drop. However, pressure drop increases with thicker fin thicknesses.

In summary, optimizing tube geometry orientation, vortex generator implementation, and fin configurations are vital strategies for enhancing heat transfer efficiency in cross-flow heat exchangers. Each approach offers unique advantages and considerations, influencing overall thermal performance and pressure drop characteristics.

Acknowledgment

The author acknowledges the use of ChatGPT 3.5 for language polishing of the manuscript.

Author details

Arash Mahdavi
Babol Noshirvani University of Technology, Babol, Iran

*Address all correspondence to: arashmahdavi93@gmail.com

References

[1] Sekulic DP, Shah RK. Fundamentals of heat exchanger design. John Wiley & Sons; 14 Nov 2023

[2] Tavousi E, Perera N, Flynn D, Hasan R. Heat transfer and fluid flow characteristics of the passive method in double tube heat exchangers: A critical review. International Journal of Thermofluids. 2023;**17**:100282

[3] Sundar LS. Synthesis and characterization of hybrid nanofluids and their usage in different heat exchangers for an improved heat transfer rates: A critical review. Engineering Science and Technology, an International Journal. 2023;**44**:101468

[4] Pourahmad S, Pesteei SM, Ravaeei H, Khorasani S. Experimental study of heat transfer and pressure drop analysis of the air/water two-phase flow in a double tube heat exchanger equipped with dual twisted tape turbulator: Simultaneous usage of active and passive methods. Journal of Energy Storage. 2021;**44**:103408

[5] Mohammadzadeh AM, Jafari B, Hosseinzadeh K. Comprehensive numerical investigation of the effect of various baffle design and baffle spacing on a shell and tube heat exchanger. Applied Thermal Engineering. 4 May 2024;**249**:123305

[6] Passoni S, Ferrario A, Ricotti ME, Mereu R. Experimental investigation of two-phase flow in Chevron-type compact plate heat exchangers: A study on pressure drops and flow regimes visualization. Applied Thermal Engineering. 2024;**242**:122542

[7] Moosavi A, Ljung AL, Lundström TS. A comparative study on thermo-fluid characteristics of free and wall-bounded cross-flow heat exchangers. Thermal Science and Engineering Progress. 2023;**40**:101746

[8] Aiba S, Tsuchida H, Ota T. Heat transfer around tubes in staggered tube banks. Bulletin of JSME. 1982;**25**(204):927-933

[9] Žukauskas A. Heat transfer from tubes in crossflow. Advances in Heat Transfer. 1972;**8**:93-160

[10] Ali MR, Al-Khaled K, Hussain M, Labidi T, Khan SU, Kolsi L, et al. Effect of design parameters on passive control of heat transfer enhancement phenomenon in heat exchangers–A brief review. Case Studies in Thermal Engineering. 2023;**43**:102674

[11] Wijayanta AT, Yaningsih I, Aziz M, Miyazaki T, Koyama S. Double-sided delta-wing tape inserts to enhance convective heat transfer and fluid flow characteristics of a double-pipe heat exchanger. Applied Thermal Engineering. 2018;**145**:27-37

[12] Zhao L, Qian Z, Wang X, Wang Q, Li C, Zhang Z. Analysis of the thermal improvement of plate fin-tube heat exchanger with straight and curved rectangular winglet vortex generators. Case Studies in Thermal Engineering. 2023;**51**:103612

[13] Tang S, Ding L, Wu X, Zhou J, Wang L, Yu Y. Numerical investigation of thermal-hydraulic characteristics in cross-flow heat exchangers with different twisted oval tubes. Case Studies in Thermal Engineering. 2024;**54**:104063

[14] Abolfathi S, Lavasani AM, Mobedi P, Afshar KS. Experimental study on flow around a tube in mixed tube bundles comprising cam-shaped and circular cylinders in in-line arrangement.

International Journal of Thermal Sciences. 2021;**163**:106812

[15] Gholami A, Wahid MA, Mohammed HA. Thermal–hydraulic performance of fin-and-oval tube compact heat exchangers with innovative design of corrugated fin patterns. International Journal of Heat and Mass Transfer. 2017;**106**:573-592

[16] Mangrulkar CK, Dhoble AS, Deshmukh AR, Mandavgane SA. Numerical investigation of heat transfer and friction factor characteristics from in-line cam shaped tube bank in crossflow. Applied Thermal Engineering. 2017;**110**:521-538

[17] Buyruk ER. Numerical study of heat transfer characteristics on tandem cylinders, inline and staggered tube banks in cross-flow of air. International Communications in Heat and Mass Transfer. 2002;**29**(3):355-366

[18] Merker GP. Heat transfer and pressure drop on the shell-side of tube-banks having oval-shaped tubes. International Journal of Heat and Mass Transfer. 1986;**29**(12):1903-1909

[19] Aiba S. Heat transfer around a tube in in-line tube banks near a plane wall. ASME Journal of Heat and Mass Transfer. 1990;**112**(4):933-938

[20] Toolthaisong S, Kasayapanand NJ. Effect of attack angles on air side thermal and pressure drop of the cross flow heat exchangers with staggered tube arrangement. Energy Procedia. 2013;**34**:417-429

[21] Horvat A, Mavko B. Heat transfer conditions in flow across a bundle of cylindrical and ellipsoidal tubes. Numerical Heat Transfer, Part A: Applications. 2006;**49**(7):699-715

[22] Berbish NS. Heat transfer and flow behavior around four staggered elliptic cylinders in cross flow. Heat and Mass Transfer. 2011;**47**:287-300

[23] Xu L, Li J, Sun K, Xi L, Gao J, Li Y. Heat analysis and optimal design of fin-and-tube heat exchanger with X-truss vortex generators. International Communications in Heat and Mass Transfer. 2024;**152**:107248

[24] Torii K, Kwak KM, Nishino K. Heat transfer enhancement accompanying pressure-loss reduction with winglet-type vortex generators for fin-tube heat exchangers. International Journal of Heat and Mass Transfer. 2002;**45**(18):3795-3801

[25] Saini P, Dhar A, Powar S. Performance enhancement of fin and tube heat exchanger employing curved delta winglet vortex generator with circular punched holes. International Journal of Thermofluids. 2023;**20**:100452

[26] Xie C, Yan G, Ma Q, Elmasry Y, Singh PK, Algelany AM, et al. Flow and heat transfer optimization of a fin-tube heat exchanger with vortex generators using response surface methodology and artificial neural network. Case Studies in Thermal Engineering. 2022;**39**:102445

[27] Chen Y, Fiebig M, Mitra NK. Conjugate heat transfer of a finned oval tube part A: Flow patterns. Numerical Heat Transfer, Part A Applications. 1998;**33**(4):371-385

[28] Sparrow EM, Kang SS. Longitudinally-finned cross-flow tube banks and their heat transfer and pressure drop characteristics. International Journal of heat and Mass Transfer. 1985;**28**(2):339-350

[29] Martinez E, Vicente W, Soto G, Salinas M. Comparative analysis of heat transfer and pressure drop in helically segmented finned tube heat exchangers. Applied Thermal Engineering. 2010;**30**(11-12):1470-1476

Chapter 4

Fundamentals and Semi-Analytical Analysis for Modeling the Capacity of Shell and Tube Heat Exchangers for Temperature Control in Hydrological Applications

Juan Gregorio Hortelano-Capetillo,
Jorge Sergio Téllez-Martínez, José Luis Zúñiga-Cerroblanco,
Alberto Saldaña-Robles, Julio César González-Juárez
and Carlos Alberto González-Rodríguez

Abstract

Currently, shell and tube heat exchangers comprise a group of equipment called profitable interest due to their usefulness in various industrial sectors. Its appropriate design is coupled to processes in a sustainable development scenario, mainly for use in laboratories or home services. The semi-analytical analysis of these complex systems currently allows for achieving temperature control objectives in the substances used, such as water, without detailed modeling of the hydrodynamics that develop due to the flow conditions. For this purpose, the development of computational tools that implement algorithms based on correlations for determining Nusselt numbers is proposed. Specifying key correlations that determine optimal operating conditions can be used to define process efficiency and contribute to the concept of net zero energy. Simultaneously, the transfer of thermal energy can continue in subprocesses, contributing to reuse and reducing the carbon footprint if the source of the energy comes from fuels or to efficiency if it comes from renewable energies.

Keywords: heat exchangers, thermal performance, recuperators simulation, hydrological applications, convective heat transfer correlations

1. Introduction

Heat exchangers are the focus of our discussion, and they are not only just essential equipment for energy transfer between fluids at different temperatures but also widely used in various industrial sectors. Their diverse construction and application in heat transfer processes have made them a staple in the petrochemical, chemical, food, and energy industries. The industry's increasing energy needs have spurred the development of cutting-edge technologies for heat exchanger manufacturing, all adhering to the rigorous standards of the Tubular Exchanger Manufacturers Association (TEMA). These standards, tailored to the class of exchanger, meticulously detail materials, design techniques, dimensions, and construction methods, ensuring optimal mechanical and thermal performance and energy-saving processes.

Heat exchangers can be classified by direct or indirect transfer [1]. In general, in an indirect heat exchanger, the heat transfer process takes place through a conductive wall that separates two fluids. In a direct heat exchanger, the fluids interact directly with each other. In the development of this chapter, the indirect heat exchanger called the shell and tube type is of particular interest. Designing various models based on parallel or countercurrent flow in a perimeter shell structure over diverse tube banks is possible. If the fluid circulates at least twice about the axial axis of the exchanger shell, it is defined as a "step" design in which the installation of internal platforms establishes recirculation.

Shell and tube heat exchangers are considered one of the most common equipment due to their ability to adapt to various processes. Therefore, various references in thermal solutions such as Sterling in the UK, Komax-Systems in the USA, HRS in India, Goje in China, and Kelvion in Brazil, among others worldwide, offer custom design services to reduce the footprint of carbon from specific processes. Reduction in greenhouse gas emissions can be achieved by adopting cost-effective technologies such as shell and tube heat exchangers. Energy efficiency is estimated to represent more than 40% of emission reductions, according to the International Energy Agency (IEA). In this way, by optimizing the processes that implement heat exchangers and using the thermal load obtained in a subprocess, a means of use that contributes to the target percentage is established.

Regarding the concept of renewable energy, currently, exchangers provide heat obtained from ambient air and geothermal water to heating networks. They also transfer heat from solar energy transformation to heat drinking water for residences and the manufacturing industry. Erdogan et al. [2] proposed the design of a shell and tube heat exchanger in combination with a parabolic trough solar collector and a geothermal power plant based on an organic Rankine cycle. The researchers developed an algorithm using the logarithmic mean temperature difference method. They coded it using ESS to determine that the baffle spacing is the critical parameter for controlling the irradiated oil temperature. They also determined that the transfer surface area in the exchanger must increase depending on the increase in solar irradiation intensity. In another investigation, Rashidi et al. [3] discussed the decrease in entropy generation and exergy analysis in shell and tube heat exchangers, determining the relevance of mass flow rates, temperatures of streams in the process, and thermal properties of fluids in the entropy generation.

Looking ahead, the optimization and design of heat exchangers are paving the way for profitable technologies that harness thermal energy efficiently in a sustainable development scenario. As the world shifts toward renewable energy sources, shell and tube heat exchangers are at the forefront of this transition. They are being integrated into the concept of net zero energy, where thermal efficiency is hailed as the first fuel. This integration not only underscores the adaptability and versatility of these heat exchangers but also inspires a vision of a future powered by sustainable and efficient energy solutions. Therefore, in the context of this chapter, some systems that implement shell and tube heat exchangers will be analyzed, adopting a methodology similar

DOI: http://dx.doi.org/10.5772/intechopen.1005676

to Erdogan et al. [2]. In the analysis, published and experimental data will be presented to implement algorithms that allow estimating the magnitudes of the temperatures of the fluids, where at least one is water, according to established processing conditions. The results would allow adjustments to the processes or the design of subprocesses for thermal energy using the fluid external to the primary process.

2. Shell and tube heat exchangers

Shell and tube heat exchangers are characterized by their design to obtain large heat transmission surfaces associated with the various tubes mounted in parallel inside a shell [4]. Due to the size of these components, the equipment can only operate with certain limitations, taking into account large pressure drops and their effect on the temperature of the fluids [5]. **Figure 1** schematically shows a shell and straight-tube heat exchanger with countercurrent fluid flow.

Shell and tube heat exchangers operate through the phenomenon of transferring thermal energy between fluids indirectly. Therefore, the greater the temperature gradient between the fluids, the greater the heat flow in transit. The primary heat transfer mechanism is convection. Therefore, since this is a complex phenomenon, the result of a specific convective condition is parameterized with data called heat transfer coefficients. In this way, by implementing these data to a thermal energy balance of the exchanger, it is possible to calculate the resulting temperature in the fluids. The energy balance can be proposed with microscopic or macroscopic models.

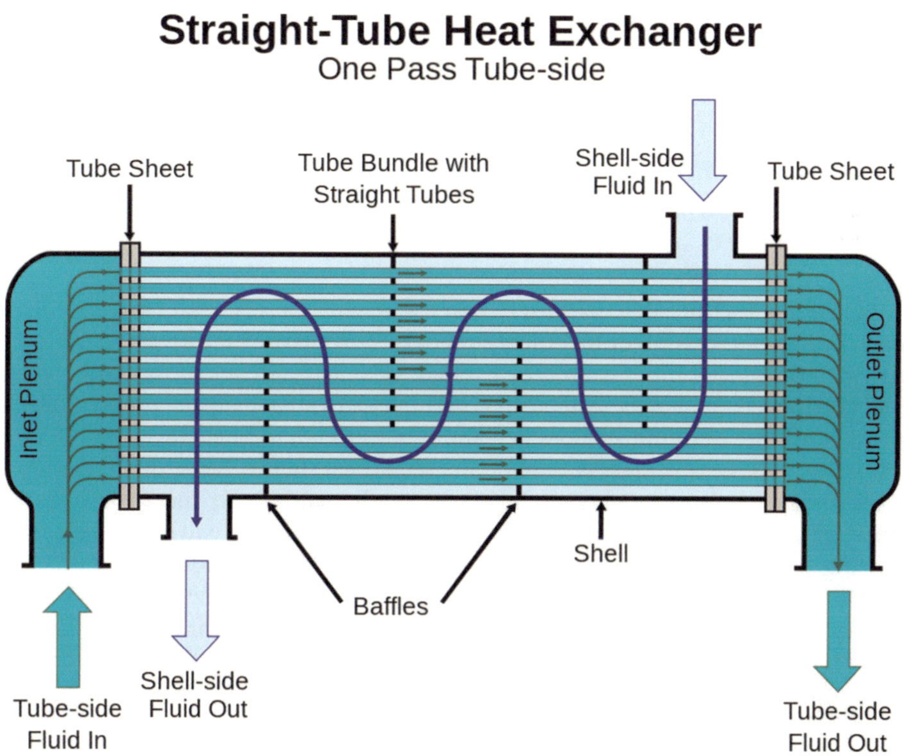

Figure 1.
Shell-and-tube type heat exchanger with countercurrent fluid flow [6].

2.1 Microscopic analysis model

The variation equations for non-isothermal systems are complex in their argumentation and, therefore, in their solution. Eqs. (1) and (2) should be solved together to obtain a distribution field of temperature profiles linked to the velocity profile field of each fluid intervening in the heat exchanger and the temperature profiles in the solids with which they interact [7, 8]. Numerical methods are required to obtain a solution and a computational system with a processing unit that performs well. Even so, the continuity equation (Eq. (1)) represents the variation in density for a fixed point as a consequence of the variations of the mass velocity vector (ρv), and the energy equation as a function of the energy flux densities and momentum (Eq. (2)) is not complete. Implementing turbulent flow regime models is needed as a characteristic of the fluid circulation in the complex exchanger.

The flow regime of the fluid circulating on the shell side (exterior of the tubes) can be justified as turbulent due to the interaction characteristics with the tubes' arrangement in its path. Furthermore, the flow direction changes concerning the location of the baffles and walls of the case, as schematically represented by the blue line in **Figure 1**. The laminar flow regime is more likely to develop in the circulation within the tubes. However, it will depend on the characteristics of the volumetric flow and the dimensions of the tubes. Under operating conditions, it would be desirable to maintain a turbulent regime to optimize the convection heat transfer mechanism and consequently increase the efficiency.

$$\frac{\partial \rho}{\partial t} = -(\nabla.\rho v) \tag{1}$$

$$\begin{aligned} \rho C_v\left(\frac{\partial T}{\partial t} + v_x\frac{\partial T}{\partial x} + v_y\frac{\partial T}{\partial y} + v_z\frac{\partial T}{\partial z}\right) &= -\left[\frac{\partial q_x}{\partial x} + \frac{\partial q_y}{\partial y} + \frac{\partial q_z}{\partial z}\right] \\ -T\left(\frac{\partial p}{\partial T}\right)_\rho\left(\frac{\partial v_x}{\partial x} + \frac{\partial v_y}{\partial x} + \frac{\partial v_z}{\partial x}\right) &- \left\{\tau_{xx}\frac{\partial v_x}{\partial x} + \tau_{yy}\frac{\partial v_y}{\partial y} + \tau_{zz}\frac{\partial v_z}{\partial z}\right\} \\ -\left\{\tau_{xy}\left(\frac{\partial v_x}{\partial y} + \frac{\partial v_y}{\partial x}\right) + \tau_{xz}\left(\frac{\partial v_z}{\partial z} + \frac{\partial v_z}{\partial x}\right) + \tau_{yz}\left(\frac{\partial v_y}{\partial z} + \frac{\partial v_z}{\partial y}\right)\right\} \end{aligned} \tag{2}$$

Calculating the fluid flow pattern due to the interaction with solid sections determines the implementation of turbulence models. Turbulence or vortex generation starts with flow instabilities caused by average velocity gradients. Consequently, the initial vortices generate new instabilities, giving rise to smaller-scale vortices. The process continues until the vortices become small enough to be nullified by the effect of viscosity, and the turbulent energy dissipates as heat. In the ducts, the fluid may not be able to develop due to the dimensions and rough quality of its surfaces, so it circulates in a turbulent regime. The use of a turbulence model will depend on analyzing the system characteristics. The equations of the κ-ε model of turbulent kinetic energy (Eq. (3)) and the turbulent dissipation rate (Eq. (4)) for analyzing the velocity profile generate acceptable results in various applications. Eqs. (3) and (4) represent the corresponding general three-dimensional formulation.

$$\frac{\partial}{\partial t}(\rho\kappa) + \frac{\partial}{\partial x_i}(\rho\kappa u_i) = \frac{\partial}{\partial x_j}\left[\frac{\mu_t}{\sigma_\kappa}\frac{\partial \kappa}{\partial x_j}\right] + 2\mu_t E_{ij}E_{ij} - \rho\varepsilon \tag{3}$$

$$\frac{\partial(\rho\varepsilon)}{\partial t}+\frac{\partial(\rho\varepsilon u_i)}{\partial x_i}=\frac{\partial}{\partial x_j}\left[\frac{\mu_t}{\sigma_\varepsilon}\frac{\partial\varepsilon}{\partial x_j}\right]+C_{1\varepsilon}\frac{\varepsilon}{\kappa}2\mu_t E_{ij}E_{ij}-C_{2\varepsilon}\rho\frac{\varepsilon^2}{\kappa} \tag{4}$$

The term E_{ij} represents the deformation rate components, u_i the velocity component in the corresponding direction and μ_t the Eddy viscosity defined by Eq. (5).

$$\mu_t=\rho C_\mu\frac{\kappa^2}{\varepsilon} \tag{5}$$

Eqs. (3) and (4) contain predefined constants that determine conditions for the number of iterations in the fitting data of the wide turbulence intervals. Some suggested data are expressed in Eq. (6).

$$C_\mu=0.09, \sigma_\kappa=1.0, \sigma_\varepsilon=1.3, C_{1\varepsilon}=1.44, C_{2\varepsilon}=1.92 \tag{6}$$

According to the fluid-structure interaction conditions, Eqs. (3) and (4) depend on the relationship between the effects of laminar and turbulent flow in a small fraction of fluid on the surface of a solid wall. In other words, the effect will depend on the dimension of the space or control volume size chosen (cell) for microscopic analysis adjacent to the surface of the solid. The dimensionless variable Y-plus or y^+ is critical to this analysis. It is calculated as a dimensionless magnitude representing the normal distance from the solid surface's wall to the cell's centre. This variable, Y-plus, evaluates the proportionality of the mesh elements' size over the solid's wall. In this way, Eqs. (7)-(10) are required to define the wall law that complements the near-wall treatment of the turbulence model (refer to **Table 1**).

$$y=\frac{y^+\mu_f}{\rho u_\tau} \tag{7}$$

$$u_\tau=\sqrt{\frac{\tau_w}{\rho}} \tag{8}$$

$$\tau_w=0.5C_f\rho U_\infty^2 \tag{9}$$

$$C_f=\frac{0.26}{Re^{1/7}} \tag{10}$$

According to the above, not all engineering problems associated with heat exchangers can be solved by equations based on laws or balances of matter, energy, and momentum because they can be very complex and involve many variables. Consequently, empirical equations determined through controlled experiments in

Turbulence model	Near wall treatment	Range of Y-plus
$\kappa-\varepsilon$	Standard wall functions	$30 < y^+ < 300$
	Non-equilibrium fall functions	$30 < y^+ < 300$
	Scallable wall functions	$y^+ > 11.225$
	Enhanced wall treatment	$y^+ < 5$

Table 1.
Y-plus specification according to the near-wall treatment of the turbulence model.

physical models were developed and adopted. The strategy consists of grouping the variables into a new dimensionless pseudo-variable to simplify the analysis. Buckingham's Pi (π) Theorem allows us to obtain, through an appropriate approach, such relationships with dimensionless numbers from a set of variables associated with a particular problem. The dimensionless numbers are identified as π-groups and determine the structure of the correlations between the transcendent variables of a physical problem [9].

As a result of the dimensional analysis, semi-analytical formulations have been developed for the phenomenological characterization of heat exchange processes for various devices, including shell and tube devices. A certain degree of empiricism determines the formulations because the analysis is implemented on the mechanical behavior of the complete system (macroscopic analysis) and not a differential element as in the microscopic model. However, its development complies with mathematical formality as long as it is limited to the interval values defined for each variable or dimensionless number.

Conditions such as reducing thermal stress, preventing leaks, and making maintenance easy due to corrosion phenomena determine that the variables of most significant interest are operating temperatures and pressures. In predicting the exchanger's performance, the heat transfer capacity is related to the overall transfer coefficient. The calculation of the global coefficient depends on the coefficients called internal and external transfer coefficients. The internal coefficient is estimated to determine the heat extraction or absorption capacity by the fluid circulating in the tubes. Respectively, the external coefficient is calculated to determine the same capacity of the fluid circulating on the shell side.

2.2 Macroscopic model: semi-analytical analysis

Even in such a structurally complex device, the heat exchange process can be defined macroscopically as a system of two substances, one giving off heat and the other absorbing it. Therefore, a global concept of the thermal phenomenon for modeling the exchanger is governed by a heat flow balance between the fluid on the shell side and that circulating on the tube side. The conditions under which this process develops can be diverse, and among the most relevant are the changes in the physical state of the substances and temporal continuity. Although it is not a general rule for applications, most heat exchangers are continuously operating systems, so they are said to reach a stable state of operation (unchanged once they reach their working configuration). Furthermore, in various systems in which water is used as the only or one of the substances, it is required to preserve its liquid state, so the formation of vapor or ice does not need to be considered. In this way, the heat flow balance is simplified and is determined by Eq. (11).

$$Q = \dot{m}_s C_s \left(\overline{T}_{es} - \overline{T}_{is}\right) = \dot{m}_t C_t \left(\overline{T}_{et} - \overline{T}_{it}\right) \tag{11}$$

Defining $\dot{m}$ as the mass flow of the substances, C as their heat capacity, and $\overline{T}$ as average temperatures. The subscripts, s and t, correspond to the specifications of the substances in the shell and tube sections, respectively. Additionally, the subscripts, i and e, signify the entry and exit locations of the fluids in the exchanger.

Given the system's complexity, considering significant mass flows of substances and turbulent flow regimes due to interaction with structures, the implementation and development of dimensional analysis and the concept of dimensionless numbers by physical-experimental modeling have represented a source of answers to acceptable control conditions.

Two methods in the analysis approach acquired relevance due to the results obtained in their application on various systems. Notably, the logarithmic mean temperature difference methods and the effectiveness-number of transfer units (ε-NTU) method can be coupled with correlations to determine convective heat transfer coefficients and thereby determine the average temperatures of the fluids in the outlet currents of the exchanger.

The following subsections address a brief description of each method's thermal calculations, a model for determining the pressure drop in fluids, correlation information supported by dimensionless numbers, and a methodology for generating a computational graphical user interface for analysis assisted by the EES engineering application, which has an efficient system of equations solver.

2.2.1 Logarithmic mean temperature difference method

Because the fluids moving along the conduits in the heat exchanger lose or gain a temperature index locally, one way to obtain a representative temperature value is to implement the logarithmic mean temperature difference concept. Eq. (12) represents a relationship for fluid flow in a countercurrent system, where T_h indicates the highest temperature of the fluid and T_c is the lowest temperature. Indices 1 and 2 represent the data concerning the fluid on the shell and tube sides, respectively [1].

$$\Delta T_{ml} = \frac{(T_{hs} - T_{ct}) - (T_{ht} - T_{cs})}{Ln\left[\frac{T_{hs}-T_{ct}}{T_{ht}-T_{cs}}\right]} \tag{12}$$

The above allows for analyzing or predicting the performance of a heat exchanger by quantifying the transferred energy Q by estimating a heat transfer coefficient called global (U_{global}), as expressed in Eq. (13) [1].

$$Q = U_g A_T \Delta T_{ml} \tag{13}$$

The approximate calculation of the global heat transfer coefficient U_g is obtained by posing the analogy with an electrical circuit in which various resistances are integrated. Thermal resistance originates from wall conduction and internal and external convection phenomena in tubes. The thermal resistance terms associated with the analogy are expressed as addends in the denominator of Eq. (14). The heat transfer coefficients associated with the fluid circulating inside the tubes (h_i) and outside (h_o) intervene in the equation, the conductivity of the material of the tube walls (k_{tm}), and the internal diameter (D_{it}) and external diameter (D_{ot}) of the tubes, as well as the fouling factor (R_{ic}) of the tube walls due to the accumulation of impurities.

$$U_g = \frac{1}{\frac{D_{ot}}{D_{it}h_i} + \frac{R_{ic}D_{ot}}{D_{it}} + \frac{D_{ot}Ln\left(\frac{D_{ot}}{D_{it}}\right)}{2k_{tm}} + \frac{1}{h_o}} \tag{14}$$

Eq. (15) defines the total area (A_T) associated with heat transfer to approximate its magnitude as the product of the external diameter of the tubes (D_{ot}), their length (L_t), and the total number of them (N_t) [1]. In Eq. (16), the term CL acquires the value of 1.0 or 0.87 if the arrangement of the tubes allows establishing a geometric relationship between their centres of 90° and 45° (square), or 30° and 60° (triangular), respectively. On the other hand, CTP acquires the values 0.93, 0.90, and 0.85, depending on whether the number of fluid passes in the equipment is defined as 1, 2, or 3, respectively. The vertical distance between the centres of the collinear tubes defines the P_t (pitch) value.

$$A_T = \pi D_{ot} L_t N_t \tag{15}$$

$$N_t = 0.785\sqrt{\frac{CL}{CTP}\frac{D_s^2}{(PR)^2 D_{ot}^2}},\ PR = \frac{P_t}{D_{ot}} \tag{16}$$

2.2.2 The effectiveness and number of transfer units (ε-NTU) method

In the ε-NTU method, the determinations of two dimensionless numbers associated with the number of transfer units (NTU) and the heat capacity of the fluids in the exchanger (C_r) are used to calculate a third dimensionless number called effectiveness (ε), which is defined as the quotient between the heat absorbed or delivered by the fluid of lower heat capacity and the maximum heat that could be exchanged. Eqs. (17) and (18) depend on the global heat transfer coefficient U_g (Eq. (14)), the total exchange area A_T (Eq. (15)) and the minor thermal capacity of the two substances in system C_m. Eq. (19), dependent on the previous equations, is defined for systems in which there is a single fluid passage in the casing, that is, a path in a non-return direction around the tube array.

$$NTU = \frac{U_g A_T}{C_m} \tag{17}$$

$$C_r = \frac{C_h}{C_m} \tag{18}$$

$$\varepsilon = \frac{1 - exp\left(-NTU(1 - C_r)\right)}{1 - C_r\, exp\left(-NTU(1 - C_r)\right)} \tag{19}$$

If the exchanger has several steps, the effectiveness calculation is obtained from Eqs. (17), (18), (20), and (21) according to Eq. (22). The expression $(NTU)_1$ refers to the calculation of Eq. (6) considering the conditions of a step, which is subsequently magnified by multiplying by the number of steps over the arrangement of tubes N_{s-s} in the exchanger.

$$NTU = N_{s-s}(NTU)_1 \tag{20}$$

$$\varepsilon_1 = 2\left[1 + C_r - \left(1 - C_r^2\right)^{0.5}\frac{1 + exp\left(-(NTU)_1\left(1 - C_r^2\right)^{0.5}\right)}{1 - exp\left(-(NTU)_1\left(1 - C_r^2\right)^{0.5}\right)}\right]^{-1} \tag{21}$$

$$\varepsilon = \left[\left(\frac{1 - \varepsilon_1 C_r}{1 - \varepsilon_1}\right)^{N_{s-s}} - 1\right]\left[\left(\frac{1 - \varepsilon_1 C_r}{1 - \varepsilon_1}\right)^{N_{s-s}} - C_r\right]^{-1} \tag{22}$$

According to the above, the calculation of the total heat transferred is obtained with Eq. (23).

$$Q = \varepsilon C_m \left(\overline{T}_{is} - \overline{T}_{it} \right) \tag{23}$$

2.2.3 Pressure drop

The pressure drop ΔP_t in the fluid due to the circulation in the tube passages N_{s-t} is obtained from Eqs. (24) and (25) [1]. The data f is defined as the friction factor estimated from the Reynolds number considering the internal diameter of the tube as the characteristic length. The variables v and ρ represent the average velocity and density of the fluid, respectively.

$$\Delta P_t = \left(\frac{4fL_sN_{s-t}}{D_{it}} + 4N_{s-t} \right) \left(\frac{\rho v^2}{2} \right) \tag{24}$$

$$f = (0.79Ln(Re_D) - 1.64)^{-2},\ Re_D = \frac{\rho v D_{it}}{\mu} \tag{25}$$

On the other hand, the pressure drop in the fluid circulating in the casing (outside the tubes) is obtained with Eqs. (26)-(31). Again, f is defined as the friction coefficient that depends on the Reynolds number Re_s, which, in turn, depends on the mass velocity of the fluid G_s and the equivalent diameter D_e for a type of arrangement of tubes (triangular or square). Parameter E is the vertical distance of clearance between tangents on the surface of the tubes. The number of deflectors is associated with parameter N_b, and the spacing between them defines the length B.

$$\Delta P_s = \frac{fG_s^2(N_b + 1)D_s}{2\rho D_e \left(\frac{\mu}{\mu_w} \right)^{0.14}} \tag{26}$$

$$f = \exp(0.576 - 0.19Ln(Re_s)),\ Re_s = \frac{G_sD_e}{\mu} \tag{27}$$

$$G_s = \frac{\dot{m}_s}{A_s},\ A_s = \frac{D_sEB}{P_t} \equiv \left[1 - \frac{D_{ot}}{P_t} \right] D_s \left[\frac{L_s}{N_b + 1} \right] \tag{28}$$

$$D_s = 0.637\sqrt{\frac{CL}{CTP}} \left[\frac{A_T(PR)^2 D_{ot}}{L_t} \right]^{0.5} \tag{29}$$

$$D_e = \frac{4\left(\frac{P_t^2\sqrt{3}}{4} - \frac{\pi D_{ot}}{8} \right)}{\frac{\pi D_{ot}}{2}},\ triangular\ arrangement \tag{30}$$

$$D_e = \frac{4\left(P_t^2 - \frac{\pi D_{ot}^2}{4} \right)}{\pi D_{ot}},\ square\ arrangement \tag{31}$$

However, the determination of the pressure drop is limited by the range of values of the Reynolds number (Re_s) from 400 to 1.0E+06. The Reynolds number is dimensionless and represents the relationship between the inertial forces and the viscous forces determined by the flow of a fluid. Therefore, it is part of the π-group, and depending on its magnitude, it characterizes the movement of a fluid in laminar or turbulent regimes. It has been determined that the critical Reynolds number (Eq. (32)) is a function of the geometry of the fluid conduit [10]. Experimentally, it is suggested that, in the flow of the boundary layer (minimum layer of stable fluid on the surface of the conduit), with a value less than or equal to 2000, a laminar flow regime is preserved; therefore, when exceeding this magnitude, it is considered the domain of a turbulent flow [9].

$$Re = \frac{\rho v D}{\mu} \tag{32}$$

The dimensionless Prandtl number (Eq. (33)) determines the proportionality between the momentum quantity's diffusion rate and the thermal diffusivity [11]. The magnitude of the Prandtl number determines the relationship between the thicknesses of the momentum and thermal boundary layers; small values define a rapid, effective diffusion of heat concerning the fluid velocity.

$$Pr = \frac{\mu/\rho}{k/\rho C_p} \tag{33}$$

2.2.4 *Use of correlations to estimate internal and external flow heat transfer coefficients*

The dimensionless Nusselt number (Nu) of the π-group represents the critical objective of the analysis of heat exchange processes since it determines the increase in heat transfer from a surface through which a fluid circulates concerning the process in which the transfer will be carried out only by heat conduction. The Nu numbers [12] can be obtained through correlations that depend on the Reynolds and Prandtl numbers. Eq. (34) expresses that estimating the heat transfer coefficient h from the dimensionless number Nu is possible. Analysis has been proposed in heat exchanger systems, such as the theories of the methods of Donohue, Tinker, Kern, Bell-Delaware, and Willis-Johnston, among others, which are based on correlations to obtain Nusselt numbers for the heat flow fluids inside tubes and on the outside or shell.

$$Nu = \frac{hL}{k} \tag{34}$$

The internal flow correlations referenced for their applicability in the analysis are those of the researchers Colburn [13], Dittus_Boelter [14], Gnielisnski [15], Sieder_Tate, Petukhov_Kirillov [16], and Notter_Sleicher. For external flow characterization, they are those of Zukauskas [17], Kern [8], Hilpert [18], Bell_Delaware [19], Taborek [20], Churchill_Berstein, and Hausen [21], for turbulent flow without phase change. **Tables 2** and **3** list the corresponding correlations for determining internal and external heat transfer coefficients. Such correlations were used to analyze the subsequent cases presented in this chapter.

2.2.5 *Introduction to the EES (engineering equation solver) program for algorithm programming*

The EES (Engineering Equation Solver) program is a valuable tool for solving problems formulated with systems of coupled nonlinear algebraic and differential equations [22].

No.	Nusselt number correlation	Limitations
1	$Nu_D = 0.023\,Re_D^{4/5}Pr^{1/3}$	$Re \geq 10000$ $0.7 \leq Pr \leq 160$ $\frac{L}{D} \geq 10$
2	$Nu_D = 0.023\,Re_D^{4/5}Pr^{0.4}$	
3	$Nu_D = \frac{\left(\frac{f}{8}\right)[Re_D - 1000]Pr}{1+12.7\left(\frac{f}{8}\right)^{0.5}\left(Pr^{2/3}-1\right)}$ $f = (0.69\,ln\,(Re_D) - 1.64)^{-2}$	$3000 \leq Re \leq 5E6$ $0.5 \leq Pr \leq 2000$
4	$Nu_D = 0.027\,Re_D^{0.8}Pr^{1/3}\left(\frac{\mu_b}{\mu_w}\right)^{0.14}$	$Re > 10000$ $0.7 \leq Pr \leq 16700$ $\frac{L}{D} > 10$
5	$Nu_D = \frac{\left(\frac{f}{2}\right)Re_D Pr}{1.07+12.7\left(\frac{f}{2}\right)^{0.5}\left(Pr^{2/3}-1\right)}$ $f = (1.58\,ln\,(Re_D) - 3.8)^{-2}$	$Re > 2100$
6	$Nu_D = 5 + 0.015\,Re_D^{m}Pr^{n}$ $m = 0.88 - \frac{0.24}{4+Pr}$ $n = 0.33 + 0.5\,exp\,(-0.6Pr)$	$1E4 \leq Re \leq 1E6$ $0.1 \leq Pr \leq 1E4$

Authors: 1, Colburn; 2, Dittus_Boelter; 3, Gnielinski; 4, Sieder_Tate; 5, Petukhov_Kirillov; 6, Notter_Sleicher.

Table 2.
Correlations for the estimation of the heat transfer coefficient for internal flow.

No.	Nusselt number correlation	Limitations
1	$Nu_D = C\,Re_{D,max}^{m}Pr^{1/3}\left(\frac{Pr}{Pr_s}\right)^{0.25}$	$10 < Re \leq 100, C = 0.9, m = 0.4$ $100 < Re \leq 1000, C = 0.683, m = 0.466$ $1000 < Re \leq 2E5, C = 0.35, m = 0.65$
2	$Nu_D = 0.36\,Re_D^{0.55}Pr^{1/3}\left(\frac{\mu}{\mu_b}\right)^{0.14}$	$2E3 < Re > 1E6$ Re_s *defined by Ecs.17 y 18*
3	$Nu_D = C\,Re_D^{m}Pr^{1/3}$	$0.4 < Re \leq 4, C = 0.989, m = 0.33$ $4 < Re \leq 40, C = 0.911, m = 0.385$ $40 < Re \leq 4000, C = 0.683, m = 0.466$ $4000 < Re \leq 4E4, C = 0.193, m = 0.618$ $4E4 < Re \leq 4E5, C = 0.027, m = 0.805$
4	$Nu_D = J_i C_p\left(\frac{\dot{m}}{A_s}\right)\left(\frac{k}{C_p\mu}\right)^{0.66}\left(\frac{\mu}{\mu_w}\right)^{0.14}$ $J_i = 0.37\,Re_s^{-0.395}$	$1E4 < Re \leq 1E5$
5	$Nu_D = 0.2\,Re_s^{0.6}Pr_s^{0.4}$	Re_s *defined by Ecs.16 y 19 o20*
6	$Nu_D = 0.3 + \frac{0.62\,Re_D^{0.5} + Pr^{0.33}}{\left[1+\left(\frac{0.4}{Pr}\right)^{0.66}\right]^{0.25}}\left[1+\left(\frac{Re_D}{282000}\right)^{0.625}\right]^{0.8}$	
7	$Nu_D = 0.35Fa\,Re_D^{0.57}Pr^{0.31}$ $Fa = 1 + 0.1S_2 + \frac{0.34}{S_1}$	*Triangular arrangement of tubes*

Authors: 1, Zukauskas; 2, Kern; 3, Hilpert; 4, Bell_Delaware; 5, Taborek; 6, Churchill_Bernstein; 7, Hausen.

Table 3.
Correlations for the estimation of the heat transfer coefficient for external flow.

The graphical user interface allows the creation of an interactive panel in which the user controls the data entry and, in turn, receives information due to the development of calculations. Therefore, this tool was used to create work files to analyze the methods in Sections 2.21 and 2.2.2 in conjunction with the correlations in Sections 2.2.3 and 2.2.4 to solve problems raised in the shell and tube heat exchangers field.

Figure 2 shows the graphical user interface of the EES program. Heat exchanger analysis algorithms require complex programming; however, each program is built from fundamental structures. A basic example with subsequent images will show the case study generation strategy. When starting the EES program in the equations window area, it will be demonstrated that introducing assignments of constants and operations between them is done relatively quickly. Previously, the unit system must be chosen by accessing it from the Options menu, as shown in **Figure 3**.

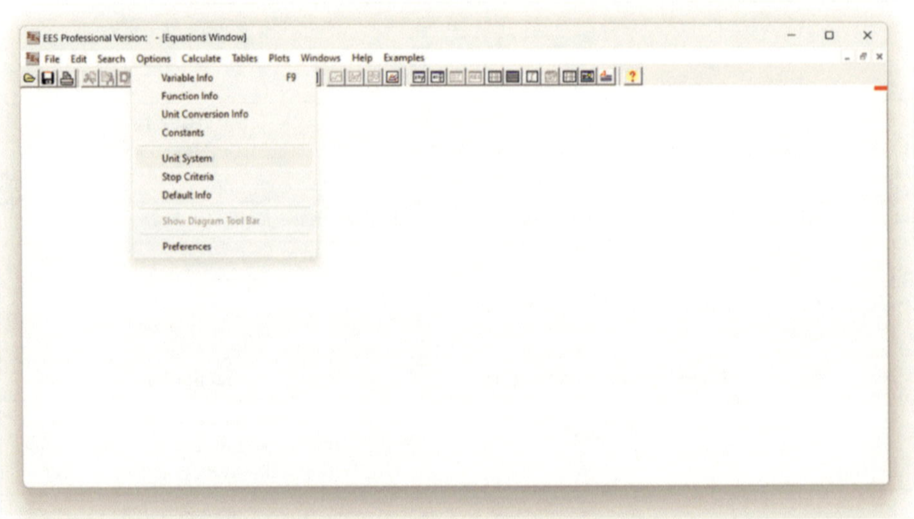

Figure 2.
The equations window is displayed initially by the graphical user interface of the EES program and access to the options of the Options menu.

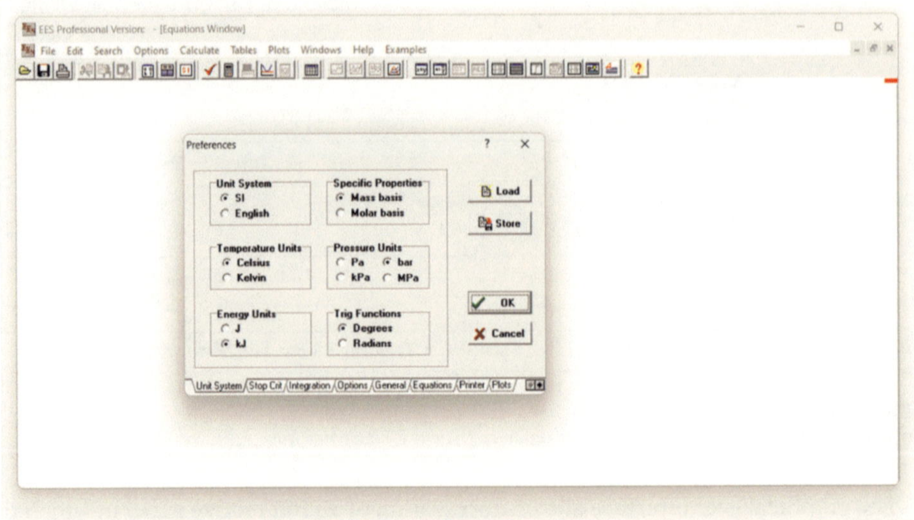

Figure 3.
The pop-up window defines the system of units in a case study.

Figure 4 shows the simple process of assigning constant values and assigning functions, such as calculating thermal capacity that depends on temperature and pressure. In turn, **Figure 5** shows the complete development in which it is now possible to estimate the magnitude of Q by activating the Solve icon. **Figure 6** shows the results window containing the information defined in the equations window and the expected result for both the Cpw and Q functions.

Alternatively, through the diagram window, it is possible to create interactive forms in which, through the toolbar, objects such as the calculation button and text

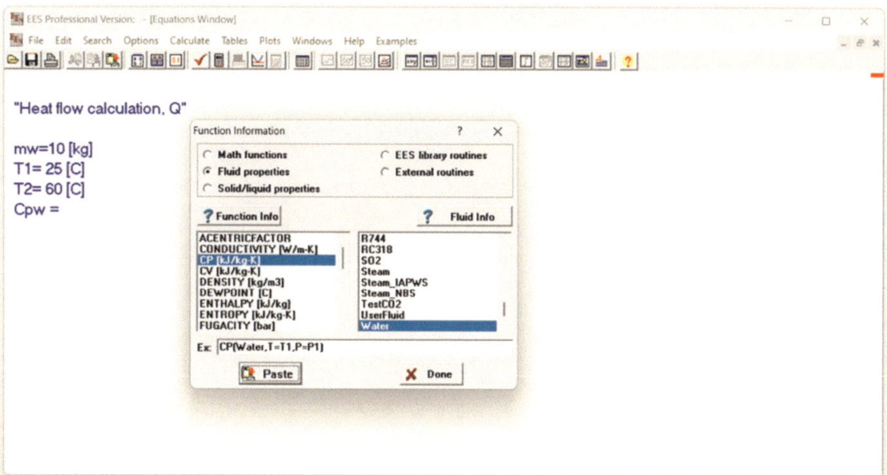

Figure 4.
Equation window with a definition of constants and a function for calculating the thermal capacity of water. Deployment of the pop-up window through the Options menu.

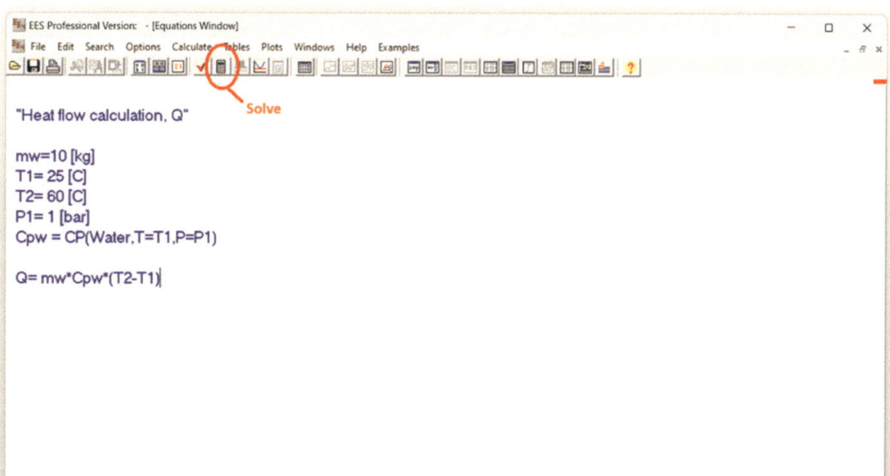

Figure 5.
The Equations window contains information about the complete calculation model that will be developed when the Solve icon is activated.

boxes linkable to the predefined constants are introduced, with which it is possible to transform them into input or output variables. **Figure 7** shows the complete process of converting the constants to variables the user can modify without accessing the equations window. Additionally, images can be integrated to understand better the analysis developed. The procedure consists of copying and pasting an edited image with the appropriate dimensions in the work area (see **Figure 8**).

However, for the conversion to be operational, the constants defined in the equation window must be converted to comments. The conversion process is simple since only the text is selected, and the secondary mouse button accesses the options pop-up to select comment{}, as shown in **Figure 9**.

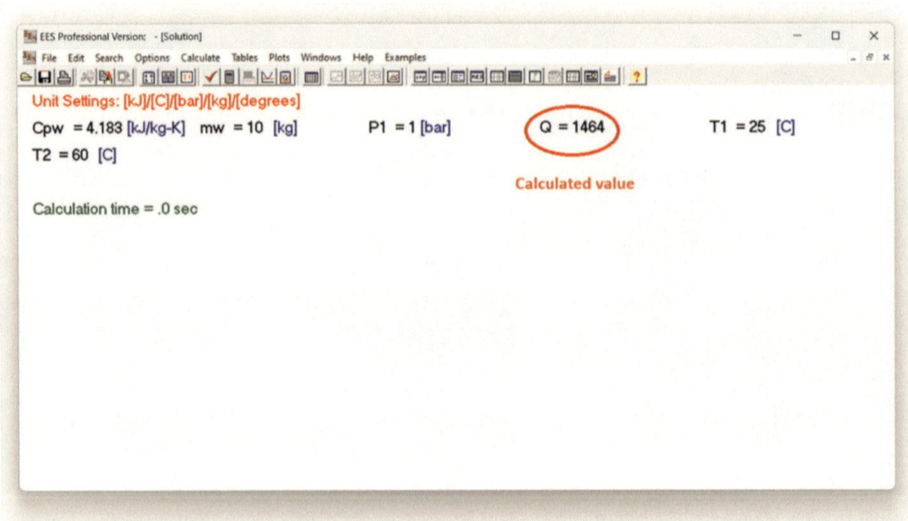

Figure 6.
Display of the results window, including the predefined data.

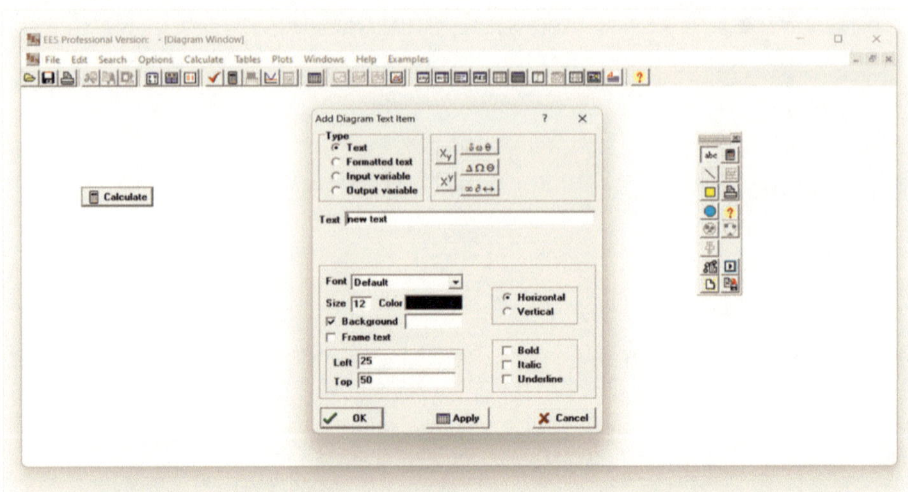

Figure 7.
Convert to variables and insert the calculation activation button using the tools available in the diagram window.

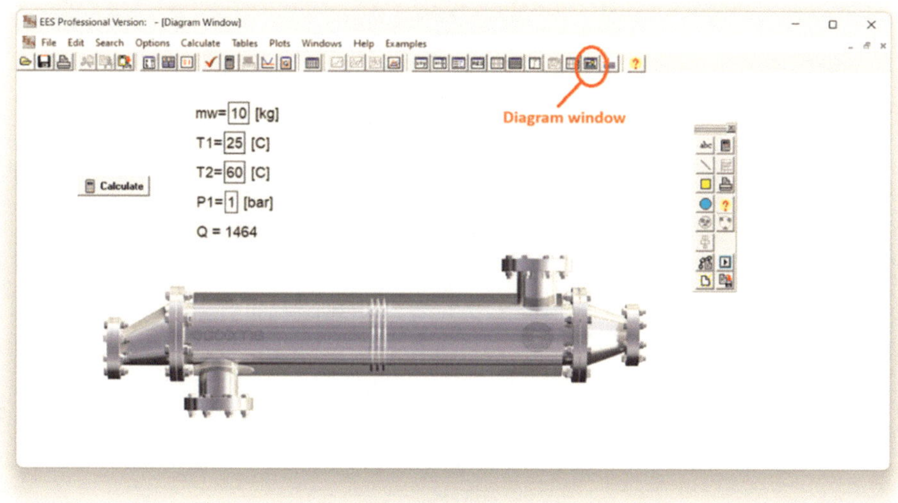

Figure 8.
Pasting schematic figures to optimize the interactive form in the diagram window.

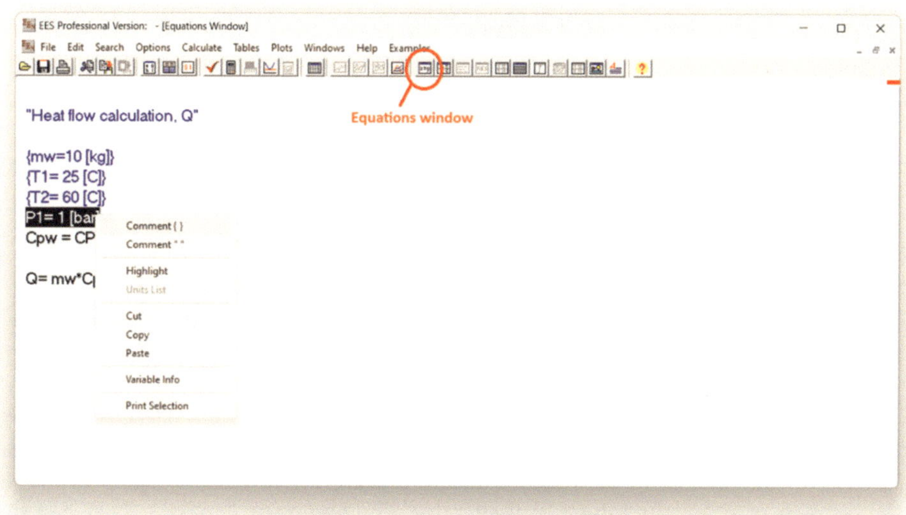

Figure 9.
Conversion of active constants to text comments through the pop-up menu displayed after selecting the text field in the equation window.

Other features can be explored in the program programming manuals. Therefore, the structure of more complex programs can be generated to solve the heat exchange process, as will be demonstrated in subsequent sections.

Figure 10 presents a general flow chart for the solution algorithms of each case study. The source quantities of the process variables are fed to each coded algorithm through each graphical user interface. Some data are processed to calculate thermal properties and subsequently transferred to a calculation cycle where the unknown dependent variables are determined. In the cycle, the systems of equations of the

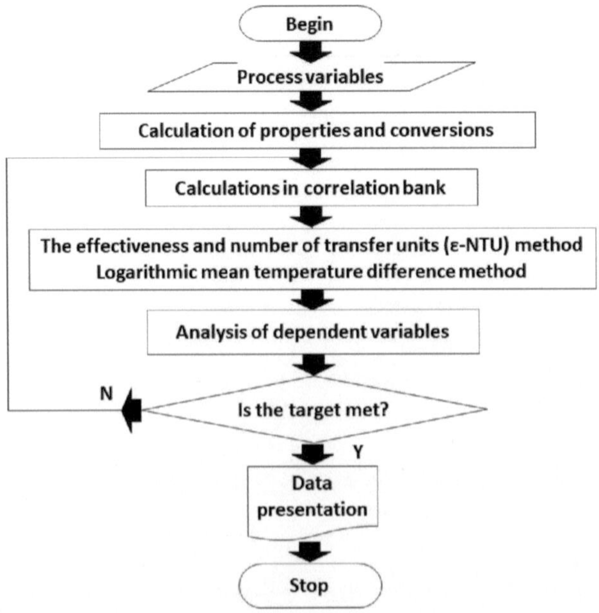

Figure 10.
Flow chart for the solution algorithms of each case study.

LMTD and $\varepsilon - NTU$ models and the correlation bank are implemented. The dependent variables are calculated with various data originating from combinations of correlations, and based on the target magnitudes, the appropriate set that characterizes the specific system is selected.

2.3 Analysis cases

2.3.1 Cylindrical exchanger for cooling water streams flowing in tubes and water streams flowing in the shell: case E1

The data on the operating conditions and geometric parameters presented in the analysis exercise were obtained from the work of Costa and Queiroz [23]. One of the objectives of the analysis of the heat exchange process is the determination of the outlet temperatures on the tube and shell sides. The comparison of data between the calculation and the source can determine that the calculation procedure proposed with the ε-NTU method constitutes a consistent calculation alternative. The operating parameters and characteristics of the heat exchanger obtained from [23] are provided in **Tables 4** and **5**. On the other hand, **Table 6** establishes the magnitudes of surface area, amount of heat, and the velocity of the fluid in the tubes.

In the analysis through the EES program, combinations of the correlations in **Tables 2** and **3** were used to determine heat transfer coefficients for internal and external flows. **Table 7** shows the magnitudes obtained from the Nutter_Sleicher and Bell_Delware correlations. With these data, the fluid temperature on the shell side closest to the target temperature was obtained with only a difference of 1.56°C. On the side of the tubes, the outlet temperature obtained was 52.27°C.

The transfer surface, the total heat flux, and the calculated fluid velocity differ slightly from the pressure drops. A high drop was determined on the tube side, and a

Section	Fluid	Volume flow	Mass flow	Inlet temp.	Pressure	Outlet temp.
		L min^{-1}	kg s^{-1}	°C	kPa	°C (target)
Shell	Water	50.18	8.33	17	12	40
Tubes	Water	84.72	13.88	67	12	—

Table 4.
Operating parameters in the cylindrical heat exchanger with water flowing on the tube and shell sides.

Section	Setting	Unit	Triangular arrangement
Shell	Length	m	0.762
	Inner diameter	m	0.438
	External diameter	m	—
	Number of steps	—	1
	Number of baffles	—	10
	Distance between baffles	m	—
Tube	Number of passages	—	4
	Distance between centers	m	0.025
	Length	m	3.048
	External diameter	m	0.03175
	Internal diameter	m	0.0301
	Amount	—	87
	Wall thermal conductivity	Wm^{-1} K^{-1}	60

Table 5.
Characteristics of the cylindrical heat exchanger with water flowing on the tube and shell sides. Case E1.

Tubes arrangement	A_T, m^2	Q, W	v, ms^{-1}
Triangular	24.63	8.13E5	1.1

Table 6.
Transfer surface characteristics and total heat flow of the cylindrical heat exchanger with specification of the velocity of the water flowing on the tube side. Case E1.

Internal flow correlations	$h_i, Wm^{-2}K^{-1}$	External flow correlations	$h_O, Wm^{-2}K^{-1}$
	Triangular arrangement		Triangular arrangement
Nutter_Sleicher	6303	*Bell_Delaware*	2098

Table 7.
The magnitudes of the heat transfer coefficients for internal and external flows gave rise to the best approximation of the shell-side outlet temperature, Case E1.

Tube arrangement	A_T, m^2	Q, W	v	$\Delta P, Pa$ (tubes)	$\Delta P, Pa$ (Shell)
Triangular	25.84	8.55E5	0.94	20,922	339.4

Table 8.
Transfer surface characteristics, total heat flow, average fluid velocity on the tube side, and pressure drops calculated for Case E1.

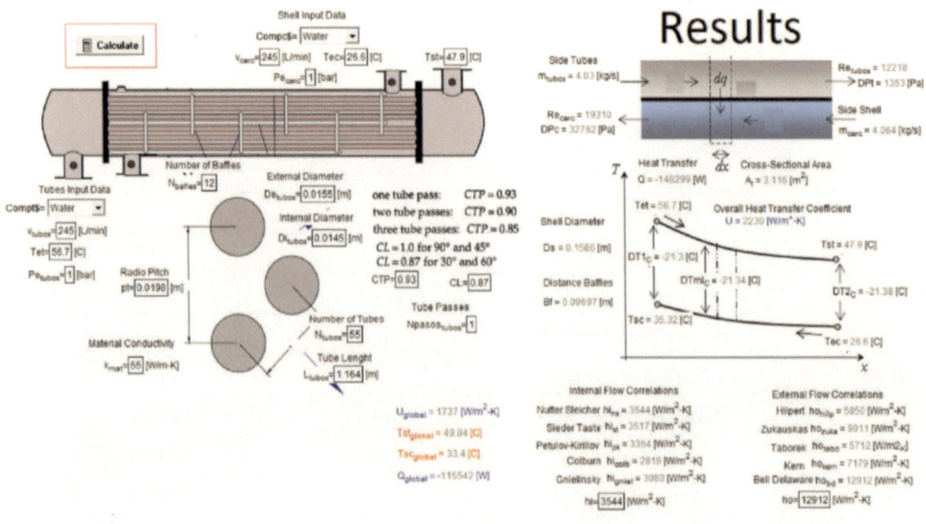

Figure 11.
An interactive form was created with the EES program for the analysis of the heat exchanger of Case E1.

shallow drop on the shell side. **Table 8** shows the results calculated using the ε-NTU method (refer to Section 2.2.2) and the pressure drop equations (refer to Section 2.2.3). **Figure 11** shows the interactive form created in the EES program for this exercise.

2.3.2 Cylindrical exchanger for cooling methanol streams flowing in the shell with water streams flowing in the tubes: Case E2

In an industrial process, methanol is cooled through a heat exchanger using water as a secondary fluid. The average shell-side methanol inlet temperature is 90°C, and the mass flow is 31.22 kg s^{-1}. The average water inlet temperature on the side of the tubes is 30°C, and the mass flow is 102.9 kg s^{-1}. The analysis aims to promote both fluids to reach an average temperature of 40°C. The operating conditions and working pressure are detailed in **Table 9**.

Table 10 provides the exchanger's operating parameters and characteristics, considering the triangular and square tube arrangement configuration.

According to the ε-NTU method (refer to Section 2.2.2) and the combinations of the correlations defined in **Tables 2** and **3**, the heat transfer coefficients for the internal and external flows presented in **Table 11** were obtained to arrange tubes with triangular and square configurations.

The transfer area, heat flow data, and efficiency were determined constant for the corresponding triangular and square configurations. Also, the pressure drop on the tube side was steady for both configurations, and the pressure drop on the shell side was only 816 Pa different. **Table 12** summarizes the corresponding information.

An analysis of combinations of correlations for calculating heat transfer coefficients for the internal (*if*) and external (*ef*) flows resulted in the values obtained for

Section	Fluid	Volume flow	Mass flow	Inlet temp.	Pressure	Outlet temp.
		L min^{-1}	kg s^{-1}	°C	kPa	°C (target)
Shell	Methanol	2485	31.22	90	1000	40
Tubes	Water	6210	102.9	30	1000	40

Table 9.
The operating parameters in the cylindrical heat exchanger are water flowing on the tube side and methanol on the shell side. Case E2.

Section	Setting	Unit	Arrangement	
			Triangular	Square
Shell	Length	m	7	7
	Inner diameter	m	0.591	0.6697
	External diameter	m	0.6	0.6757
	Number of steps	—	1	1
	Number of baffles	—	8	8
Tube	Number of passages	—	1	1
	Distance between centers	m	0.018	0.018
	Length	m	6.025	6.025
	External diameter	m	0.016	0.016
	Internal diameter	m	0.01587	0.01587
	Amount	—	748	748
	Wall thermal conductivity	Wm^{-1} K^{-1}	15.1	15.1

Table 10.
Characteristics of the cylindrical heat exchanger, with water flowing on the tube side and methanol on the shell side, with triangular and square tube arrangement configuration. Case E2.

Internal flow correlations	h_i, $Wm^{-2}K^{-1}$		External flow correlations	h_o, $Wm^{-2}K^{-1}$	
	Tube arrangement			Tube arrangement	
	Triangular	Square		Triangular	Square
Petukov	4122	4122	*Taborek*	711.9	717.9
Gnielinski	3966	3966	*Zukauskas*	1146	1178
Dittus	3739	3739	*Hilpert*	1274	1309
Colburn	3344	3344	*Kern*	1271	1438

Table 11.
Magnitudes of heat transfer coefficients for internal and external flows are calculated with various correlations for both a triangular and a square tube array configuration. Case E2.

the global heat transfer coefficient and the fluid outlet temperature on the tube and shell sides in **Tables 13** and **14**. For both cases of tube arrangement configurations, the temperatures are very close to the target value of 40°C. The furthest values are associated with combining the Petukhov_Kirillov and Taborek correlations.

By obtaining results close to the target outlet temperatures with different combinations of correlations, it is verified that the exchanger operating conditions are within its predictive domain. From this perspective, the calculation procedure implemented for the form interface in the EES program, shown in **Figure 12**, can be used to generate data curves for various systems by modifying values in **Table 9**. However, the study could also be extended by modifying the values from **Table 10**.

Tube arrangement	A_T, m^2	Q, W	ε	$\Delta P, Pa$ (tubes)	$\Delta P, Pa$ (Shell)
Triangular	227.3	4.28E6	0.83	3592	4856
Square	227.3	4.28E6	0.83	3592	5672

Table 12.
Transfer surface characteristics and total heat flow, efficiency, and pressure drops calculated for Case E2.

Case	Correlation	U_g, Wm-2 K-1	$\overline{T}_{os}$, °C	$\overline{T}_{ot}$, °C
1	*Gnielinski (if)*	881.7	40.77	39.95
	Zukauskas (ef)			
2	*Dittus-Boelter (if)*	869.8	40.92	39.92
	Zukauskas (ef)			
3	*Colburn (if)*	977.2	39.69	40.17
	Kern (ef)			
4	*Petukhov (if)*	591.6	46.42	38.81
	Taborek (ef)			
5	*Petukhov (if)*	889.2	40.67	39.97
	Zukauskas (ef)			
6	*Gnielinski (if)*	955.1	39.91	40.13
	Hilpert (ef)			
7	*Dittus-Boelter (if)*	941.3	40.06	40.1
	Hilpert (ef)			
8	*Gnielinski (if)*	1024	39.25	40.26
	Kern (ef)			
9	*Colburn (if)*	846.4	41.24	39.86
	Zukauskas (ef)			

Table 13.
Results of the overall heat transfer coefficient and fluid outlet temperatures on the tube and shell sides by combining correlations for a triangular configuration in the tubes. Case E2.

Case	Correlation	U_g, Wm-2 K-1	$\overline{T}_{os}$, °C	$\overline{T}_{ot}$, °C
1	*Gnielinski (if)*	942.3	40.1	40.05
	Zukauskas (ef)			
2	*Dittus-Boelter (if)*	928.8	40.2	40.07
	Zukauskas (ef)			
3	*Colburn (if)*	917	40.33	40.04
	Kern (ef)			
4	*Petukhov (if)*	604.7	46.42	38.88
	Taborek (ef)			
5	*Petukhov (if)*	950.9	39.96	40.12
	Zukauskas (ef)			
6	*Gnielinski (if)*	1021	39.28	40.25
	Hilpert (ef)			
7	*Dittus-Boelter (if)*	1005	39.28	40.25
	Hilpert (ef)			
8	*Gnielinski (if)*	958.5	39.88	40.13
	Kern (ef)			
9	*Colburn (if)*	902.2	40.51	40.01
	Zukauskas (ef)			

Table 14.
Results of the global heat transfer coefficient and the fluid outlet temperatures on the tube and shell sides using a combination of correlations for a square configuration in the tubes. Case E2.

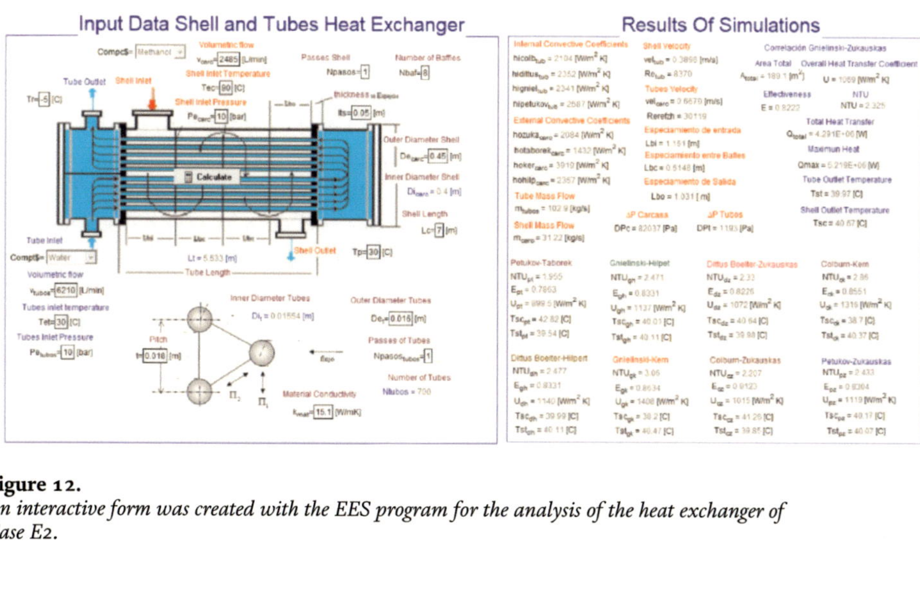

Figure 12.
An interactive form was created with the EES program for the analysis of the heat exchanger of Case E2.

2.3.3 Cylindrical exchanger for cooling oil streams flowing in the tubes and water stream flowing in the casing: case E3

A heat exchanger designed to cool oil was instrumented with thermocouples to measure the inlet and outlet temperatures of the fluids on the tube and shell sides. Processing data for 30 tests are presented in **Table 15**.

The characteristics of the heat exchanger are presented in **Table 16**.

The graph in **Figure 13(a)** shows the measured and calculated oil outlet temperature data for each test in **Table 15**. As noted, the Sider_Tate and Hilpert correlations generated results close to the data recorded by the thermocouples in the tests on the exchanger. The results of the Dittus_Boelter-Churchill_Bernstein and Dittus_Boelter-Kern combinations were eliminated because they significantly exceeded the desired temperature. In turn, the graph in **Figure 13(b)** shows the data for water. In this case, the calculations with the combination of the Sieder_Tate-Hilper, Gnielinski-Zukauskas and Sieder_Tate-Churchill_Berstein correlations adequately approximate the target temperature.

According to the above, the analysis through development in the ESS program allowed obtaining a control tool on the behavior of the heat exchanger for oil cooling. It is natural for the physical properties of the oil to change due to the recirculatory use of the oil and contamination of the oil. However, the information can be adjusted to the data required by the calculation procedure until the results of the calculated outlet temperatures converge. This condition turns the computational tool into a simulator with prediction capacity.

The form was created in the EES program for analysis using the ε-NTU method in **Figure 14**.

2.3.4 Cylindrical exchanger for cooling water streams flowing in tubes and R134a refrigerant stream flowing in the casing: case E4

An exchanger is used for zone cooling and water, and R134a refrigerant is involved. The objective of the analysis is to determine the appropriate correlation information to validate the exchanger's behavior using the ε-NTU method. **Table 17** summarizes the operation data recorded in seven processes.

The behavior of the exchanger is characterized by three regions associated with the refrigerant's aggregation state. The first zone (Z1) is associated with the decrease in the temperature of the refrigerant in the vapor state from the maximum temperature to the change of state temperature, that is, to the condensation temperature. The second zone (Z2) is associated with the transformation process by condensation to the liquid state, and the third zone (Z3) with cooling until the refrigerant's minimum temperature or outlet temperature on the shell side is reached. **Table 18** summarizes the characteristics of the exchanger with a triangular tube arrangement configuration.

The thermal processes in the three zones will be characterized according to the correlations that allow the estimation of the convective coefficients that determine the amount of heat transferred. Therefore, the approach to the energy balance and the identification of the heat exchange surfaces will support the description of the thermal phenomenon in each zone.

In Z1, the cooling of the superheated vapor that enters the condenser from the discharge of the compressor of the system in which the exchanger operates originates. At Z2, the refrigerant goes from the saturated vapor to the saturated liquid state. In this zone, condensation will be considered to occur at a constant temperature (such as a pure fluid or azeotropic mixture), so it is easy to define that the water stream has the minimum thermal capacity. Finally, in Z3, the temperature is decreased below the saturation condition. In this regard, the minimum capacitance is determined in Z1 and Z3, and the efficiency of the exchanger in Z2 is defined by Eq. (35).

Test	Oil (tube side)			Water (shell side)		
	Volume flow, Ls^{-1}	Temperature, °C		Mass flow, gs^{-1}	Temperature, °C	
		Inlet	Outlet		Inlet	Outlet
1	0.02	45	34.08	10.83	21.9	31.1
2	0.02	44	32.42	10.83	21.92	30.5
3	0.02	43	32.26	10.83	21.55	31.1
4	0.02	42	31.04	10.83	20.9	29.32
5	0.02	41	30.07	10.83	21.03	29.1
6	0.02	40	30.58	10.83	21.05	29.37
7	0.02	39	29.9	10.83	21.1	28.24
8	0.02	38	30	10.83	21.12	28.51
9	0.02	37	29.03	10.83	21.14	28
10	0.02	36	27.09	10.83	21.18	27.6
11	0.02	35	27.65	10.83	21.11	26.5
12	0.02	45	34.19	10.83	22.08	32.36
13	0.019	45	35.07	10.83	22.1	32.59
14	0.0181	45	34.02	10.83	22.13	32.83
15	0.0171	45	33.22	10.83	22.13	33
16	0.0162	45	33.48	10.83	22.18	33.3
17	0.0152	45	33.08	10.83	22.28	33
18	0.0143	45	33	10.83	22.28	29.9
19	0.0133	45	32.78	10.83	22.37	29.5
20	0.0123	45	32.6	10.83	22.34	31.2
21	0.02	45	34.02	10.83	22.38	31.1
22	0.02	45	34.03	10.5	22.41	31.5
23	0.02	45	34.35	10.1	22.43	31.8
24	0.02	45	34.46	9.8	22.46	32
25	0.02	45	34.56	9.5	22.47	32
26	0.02	45	34.82	9	22.61	32.1
27	0.02	45	34.94	8.8	22.62	32.3
28	0.02	45	35.14	8.5	22.68	32.7
29	0.02	45	35.32	8.1	22.7	33
30	0.02	45	35.42	7.8	22.73	33.6

Table 15.
Test data on the heat exchanger for oil cooling. Case E3.

$$\eta_{Z2} = 1 - exp\,(-NTU_{Z2}) \tag{35}$$

NTU_{Z1} and the global heat transfer coefficient U_g are determined with Eqs. (14) and (17), respectively. However, Eq. (14) is modified to introduce a heat transfer coefficient associated with the condensation phenomenon derived from the extension of Nusselt's analysis [10] defined by Eq. (36).

$$h_{Z2} = 0.729\left[\frac{g\rho_l(\rho_l - \rho_v)k_l^3 H'_{fg}}{N_t\mu_l(T_{sat} - T_{wall})D_{os}}\right]^{0.25} \tag{36}$$

The term H'_{fg} represents the latent heat modified by thermal advection effects that can be obtained by Eq. (37).

$$H'_{fg} = H_{fg}(1 + 0.68J_a),\ J_a = Cp_l\left(\frac{T_{sat} - T_{wall}}{H_{fg}}\right) \tag{37}$$

Section	Setting	Unit	Arrangement triangular
Shell	Length	m	0.3048
	Inner diameter	m	0.127
	External diameter	m	0.133
	Number of steps	—	1
	Number of baffles	—	9
	Wall thermal conductivity	$Wm^{-1}K^{-1}$	15.1
Tube	Number of passages	—	1
	Distance between centers	m	0.00794
	Length	m	0.3048
	External diameter	m	0.0065
	Internal diameter	m	0.0055
	Amount	—	172
	Wall thermal conductivity	$Wm^{-1}K^{-1}$	60

Table 16.
Characteristics of the cylindrical heat exchanger, with oil flowing on the tube side and water on the shell side, with triangular tube arrangement configuration. Case E3.

Eq. (38) expresses the heat exchange surfaces and the total heat transfer in the total balance.

$$A_T = A_{Z1} + A_{Z2} + A_{Z3}, Q = Q_{Z1} + Q_{Z2} + Q_{Z3} \tag{38}$$

In the analysis procedure, different correlations were combined to calculate the heat transfer coefficients for internal and external flow. In summary, the association of the Dittus_Boelter-Hilpert correlations were the most appropriate to characterize the thermal phenomena in zones Z1 and Z3. For Z2, the Dittus_Boelter-Nusselt

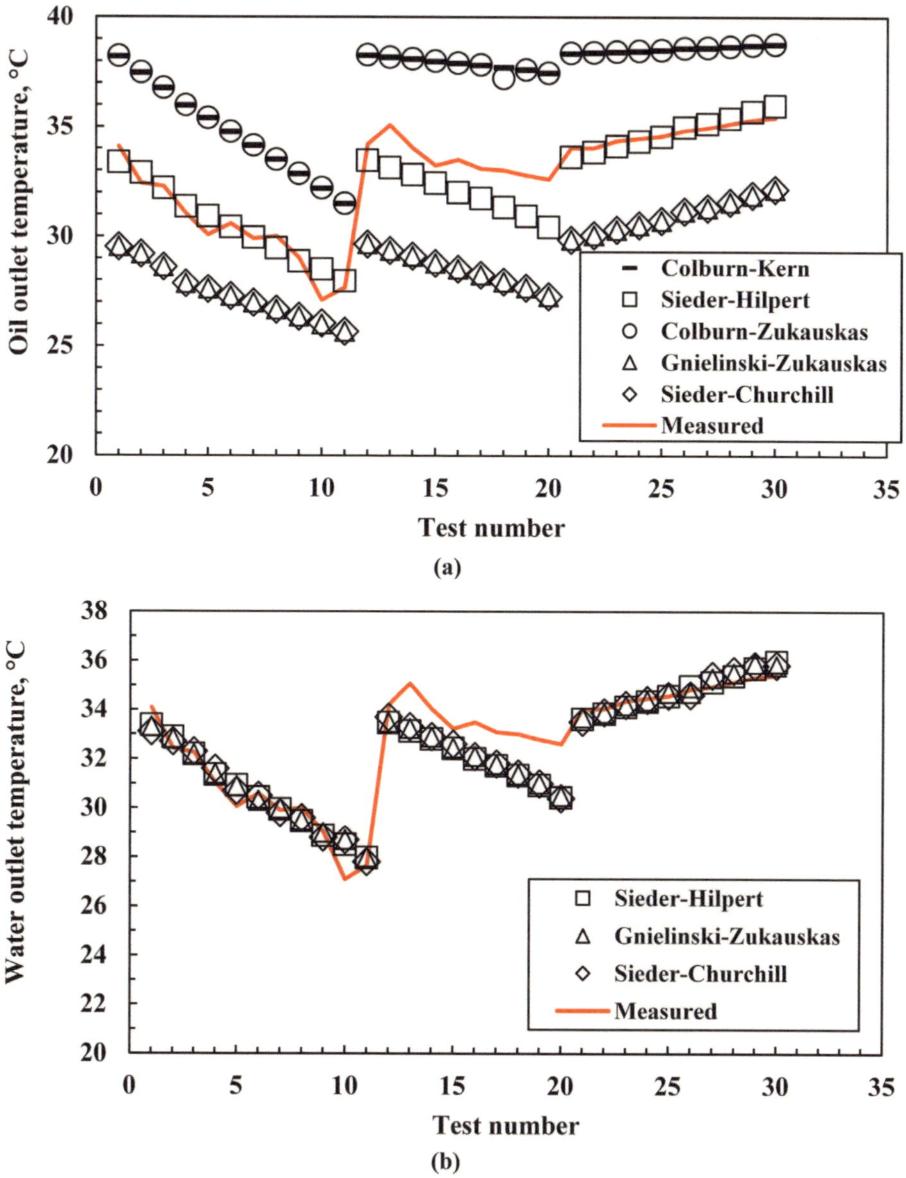

Figure 13.
Measured outlet temperatures (listed in ***Table 15****) and calculated with internal and external flow correlations for fluids: (a) oil and (b) water.*

combination was used. **Table 19** summarizes the information on the transfer surface and the amount of heat transferred calculated for each exchanger zone. Also, the corresponding effectiveness is listed in the table, which indicates a relatively regular behavior for the seven processes analyzed.

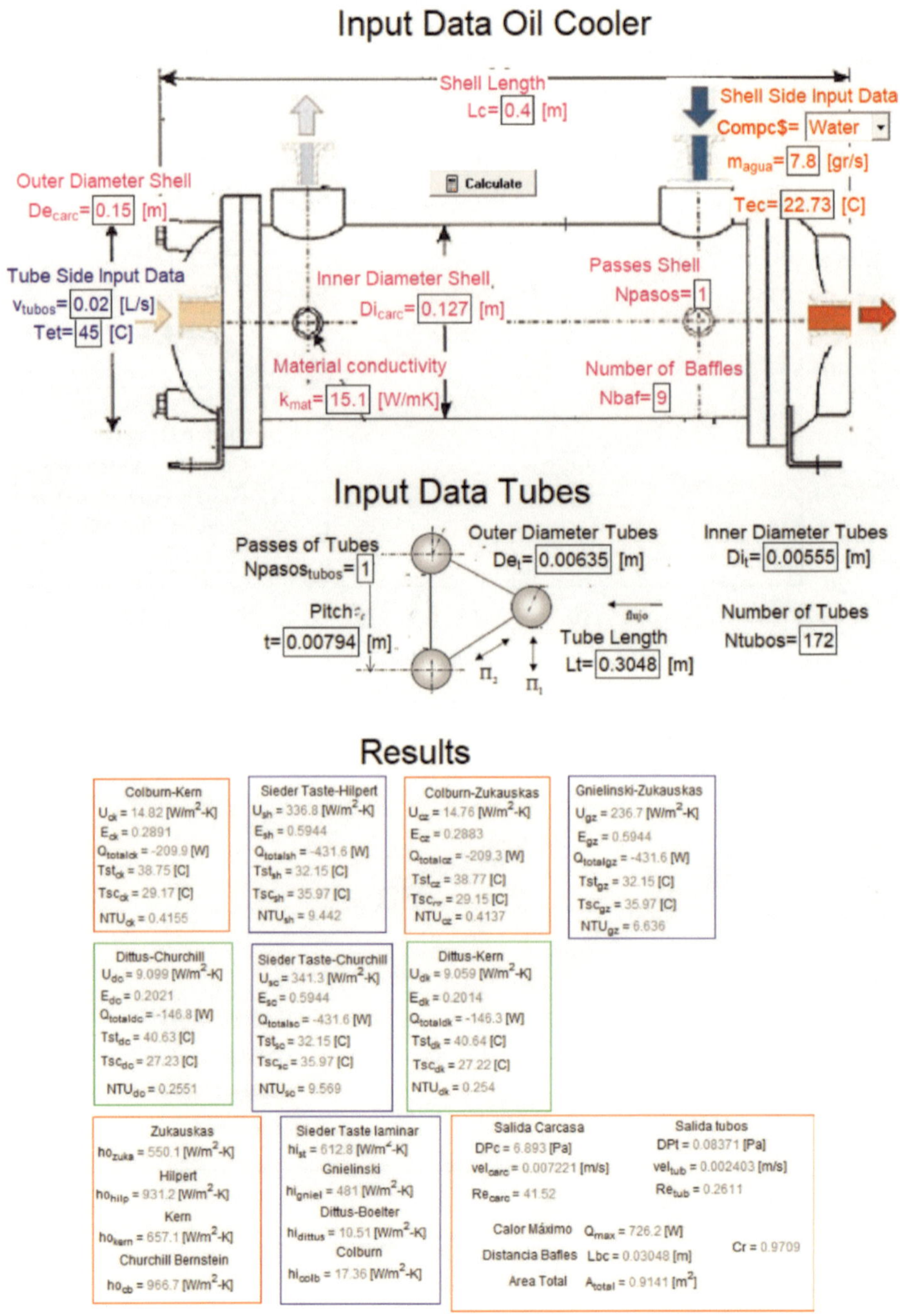

Figure 14.
An interactive form was created with the EES program for the analysis of the heat exchanger of Case E3.

Regarding the outlet temperatures of the fluids, measured and calculated on the tube and casing sides, it is observed that there is a good approximation, and the maximum differences do not exceed 3 °C. Therefore, the model can be considered

Case	Section	Volume flow	Mass flow	Inlet temp.	Pressure	Outlet temp.
		m3 h^{-1}	kg s^{-1}	°C	Bar	°C
1	Shell	—	0.055	56.5	11.36	42.68
	Tubes	1.021	—	23.83	—	32.02
2	Shell	—	0.0541	68.65	14.95	52.34
	Tubes	0.581	—	26.61	—	39.87
3	Shell	—	0.0603	66.76	13.3	47.84
	Tubes	0.588	—	15.87	—	31.04
4	Shell	—	0.0466	51.2	10.73	40.83
	Tubes	1.027	—	24.72	—	31.6
5	Shell	—	0.0551	67.98	14.58	51.18
	Tubes	0.638	—	26.06	—	38.5
6	Shell	—	0.036	45.67	9.54	36.54
	Tubes	1.042	—	23.69	—	29.17
7	Shell	—	0.0576	72.57	16.02	27.07
	Tubes	0.563	—	27.07	—	41.83

Table 17.
Measurement data in the heat exchanger for oil cooling. Case E4.

Section	Setting	Unit	Arrangement Triangular
Shell	Length	m	0.87
	Inner diameter	m	0.183
	External diameter	m	0.195
	Number of steps	—	1
	Number of baffles	—	4
Tube	Number of passages	—	2
	Distance between centers	m	0.0195
	Length	m	0.8
	External diameter	m	0.016
	Internal diameter	m	0.013
	Amount	—	10
	Wall thermal conductivity	$Wm^{-1} K^{-1}$	60

Table 18.
Characteristics of the cylindrical heat exchanger, with water flowing on the tube side and R134a refrigerant on the shell side, with triangular tube arrangement configuration. Case E4.

Case	Cooling in Z1			Condensation			Cooling Z2		
	A_{Z1}, m^2	ε	Q_{Z1}, W	A_{Z2}, m^2	ε	Q_{Z2}, W	A_{Z3}, m^2	ε	Q_{Z3}, W
1	0.0817	0.103	125.2	1.048	0.355	8716	1.74	0.752	842.7
2	0.124	0.136	235.9	0.903	0.403	7875	1.84	0.747	1123
3	0.093	0.1	235.1	0.902	0.383	9129	1.874	0.732	14.92
4	0.066	0.09	82.63	1.034	0.357	7496	1.77	0.783	623.1
5	0.13	0.14	250.6	0.898	0.384	8092	1.84	0.749	1161
6	0.065	0.11	58.1	0.977	0.346	5957	1.83	0.83	412.6
7	0.163	0.165	328.3	0.887	0.4	8171	1.82	0.73	1301

Table 19.
Transfer surface data and the amount of heat transferred were calculated for each zone, and the effectiveness obtained for each process. Case E4.

Process	Outlet temperature results, °C			
	Refrigerant R134a (side shell)		Water (side tubes)	
	Measured	Calculated	Measured	Calculated
1	56.5	54.65	32.02	31.99
2	71.11	68.65	39.87	40.3
3	69.76	66.76	31.04	31.76
4	53.6	51.2	31.6	31.59
5	69.9	67.98	38.5	38.88
6	48.3	45.67	29.17	29
7	75.2	72.57	41.83	42.05

Table 20.
Measured and calculated fluid outlet temperatures for the R134a refrigerant cooling heat exchanger processes on the shell side with water on the tube side. Case E4.

very useful for predicting exchanger performance. **Table 20** summarizes the corresponding data and allows a comparison of the information.

2.3.5 Cylindrical exchanger for cooling water streams flowing in tubes and water streams flowing in the shell: case E5

Some shell and tube-type heat exchangers are designed to heat water in swimming pools with a maximum volume of 450 m^3 through a simple and flexible installation. Hot water can come from a boiler, heat pump, and solar panel systems, among others, toward the shell side with a maximum volumetric flow of 200 L min^{-1}. Cold water enters from the side of the tubes with a maximum volumetric flow of 25 L min^{-1}. **Table 21** includes additional information on the exchanger operation and suggests three analysis processes.

Table 22 lists the characteristics of the heat exchanger models with differences in the heat transfer surface. Model 1 defines an area of 2.65 m^2, while model 2 defines 2.96 m^2.

Section	Fluid	Volume flow	Pressure	Inlet temperature		
		L min^{-1}	kPa	°C (proc. 1)	°C (proc. 2)	°C (proc. 3)
Shell	Water	200	31.22	60	70	45
Tubes	Water	25	102.9	10	20	10

Table 21.
Operating parameters in the cylindrical heat exchanger include water flowing on the tube side and water on the shell side. Case E5.

Section	Setting	Unit	Triangular arrangement	
			Model 1	Model 2
Shell	Length	m	1.2	1.6
	Inner diameter	m	—	—
	External diameter	m	—	—
	Number of steps	—	1	1
	Number of baffles	—	6	8
Tube	Number of passages	—	1	1
	Distance between centers	m	0.04	0.045
	Length	m	1.2	1.6
	External diameter	m	0.02	0.031
	Internal diameter	m	0.019	0.03
	Amount	—	37	19
	Wall thermal conductivity	Wm^{-1} K^{-1}	60	60

Table 22.
Characteristics of the cylindrical heat exchanger, with water flowing on the tube side and water on the shell side, triangular tube arrangement configuration. Case E5.

Model	Side	Water outlet temperature, °C (process 1)				
		Fluent	*Petukhov-Taborek*	*Gnielinski-Zukauskas*	*Dittus-Hausen*	*Colburn-Kern*
1	Shell	31.64	31.06	13.92	31.25	25.29
	Tubes	57.29	57.32	59.5	57.3	58.05
2	Shell	29.31	30.56	17.63	30.96	27.02
	Tubes	57.26	57.38	59.03	57.33	57.83

Table 23.
Water outlet temperatures on the tube and shell sides for the two models of the exchanger with conditions for process 1. Case 5.

In this case, in addition to the macroscopic semi-analytical analysis essentially referred to as the ε-NTU method, the results of the microscopic analysis obtained using the Ansys Fluent program are presented. **Tables 23–25** summarize the results of the water outlet temperature on the tube and shell sides for the two heat exchanger models and the three processes, both for the macroscopic analysis (ε-NTU) and for the microscopic analysis (Ansys Fluent).

Model	Side	Fluid outlet temperature, °C (process 2)				
		Fluent	*Petukhov-Taborek*	*Gnielinski-Zukauskas*	*Dittus-Hausen*	*Colburn-Kern*
M1	Shell	41	42.72	27.96	42.9	36.67
	Tubes	67.3	67.1	68.3	67.02	67.87
M2	Shell	39.5	41.42	32.72	41.6	38.8
	Tubes	67	67.14	68.38	67.08	67.6

Table 24.
Water outlet temperatures on the tube and shell sides for the two models of the exchanger with conditions for process 2. Case 5.

Model	Side	Fluid outlet temperature, °C (process 3)				
		Fluent	*Petukov-Taborek*	*Gnielinski-Zukauskas*	*Dittus-Hausen*	*Colburn-Kern*
M1	Shell	26.7	25.6	12.66	25.74	20.53
	Tubes	42.3	43.16	44.66	43.14	43.67
M2	Shell	25.64	24.31	15.3	24.66	21.85
	Tubes	43.06	43.2	44.33	43.15	43.5

Table 25.
Water outlet temperatures on the tube and shell sides for the two models of the exchanger with conditions for process 3. Case 5.

When analyzing the information in **Tables 23–25**, it is observed that the results obtained with the combinations of the Petukhov_Kirillov-Taborek and Dittus_Boelter-Hausen coefficients present the most minor difference concerning the results obtained with Ansys Fluent. Due to the increase in the heat transfer surface in model 2, the temperature is consistently lower than that obtained for model 1 in all processes.

Tables 26 and **27** summarize the results of the heat transfer coefficient calculations for the internal and external flow of each process and model, respectively. The combination of Petukhov_Kirillov-Taborek and Dittus_Boelter-Hausen coefficients generates similar results for the overall heat transfer coefficient (refer to **Table 28**).

Internal flow correlations	$h_i, Wm^{-2}K^{-1}$					
	Process 1		Process 2		Process 3	
	M1	M2	M1	M2	M1	M2
Petukhov	535.4	375.5	611.2	428.6	535.4	375
Gnielinski	72.83	104.6	208.3	193.2	73	105
Dittus	479.3	338.4	542.5	383	479.3	338.4
Colburn	409	288.8	472.7	333.8	409	288.8

Table 26.
Estimated heat transfer coefficients for the internal flow of the two exchanger models with conditions for the three processes. Case 5.

External flow correlations	$h_o, Wm^{-2}K^{-1}$					
	Process 1		Process 2		Process 3	
	M1	M2	M1	M2	M1	M2
Taborek	1475	2336	1525	2417	1388	2199
Zukauskas	248.8	1754	277.2	1958	216	1517
Hausen	2410	17,820	2517	18,615	2229	16,485
Kern	686.6	1975	717.1	2064	649.3	1869

Table 27.
Estimated heat transfer coefficients for the external flow of the two exchanger models with conditions for the three processes. Case 5.

Correlation association	$U_g, Wm^{-2}K^{-1}$					
	Process 1		Process 2		Process 3	
	M1	M2	M1	M2	M1	M2
Petukhov-Taborek	373.4	323.4	414.6	364	367.6	320.6
Gnielinski-Zukauskas	54.03	98.67	115.1	175.9	53	98
Dittus-Hausen	378	332.1	421.6	375.3	373.3	331.6
Colburn-Kern	246	252	273.6	287.3	241.1	250

Table 28.
Estimated global heat transfer coefficients of the two exchanger models with conditions for the three processes. Case 5.

Although the magnitudes of the heat transfer coefficients for the internal flow vary, they balance with the determined magnitudes of the heat transfer coefficients for the external flow. Therefore, the calculated efficiency presents the same pattern as the overall heat transfer coefficient (**Table 29**). The above indicates that macroscopic balances are certainly not precise, although they represent a solution for the system in approximation to the microscopic balance. The form was created in the EES program for analysis using the LMTD and ε-NTU methods in **Figure 15**.

On the other hand, the simulation of the process based on the microscopic balance will depend significantly on the precise definition of the system's geometric characteristics. Therefore, if the details of the shapes and the amount of information about

Correlation association	ε					
	Process 1		Process 2		Process 3	
	M1	M2	M1	M2	M1	M2
Petukov-Taborek	042	0.41	0.45	0.44	0.41	0.4
Gnielinski-Zukauskas	0.1	0.15	0.15	0.25	0.1	0.15
Dittus-Hausen	0.41	0.4	0.45	0.44	0.41	0.4
Colburn-Kern	0.33	0.34	0.33	0.37	0.3	0.33

Table 29.
Estimated efficiency of the two exchanger models with conditions for the three processes. Case 5.

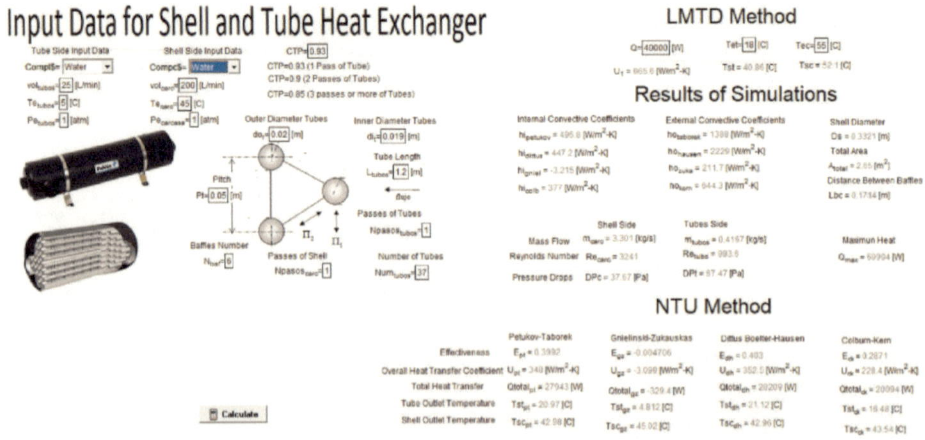

Figure 15.
An interactive form was created with the EES program for the analysis of the heat exchanger of Case E5.

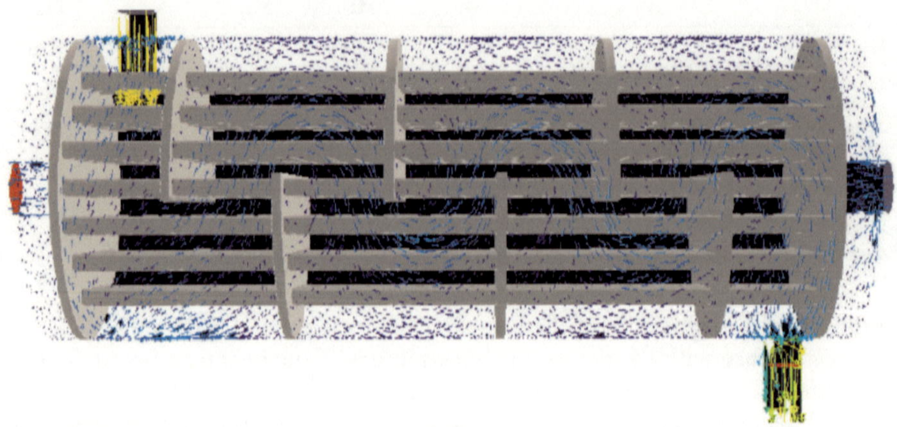

(a)

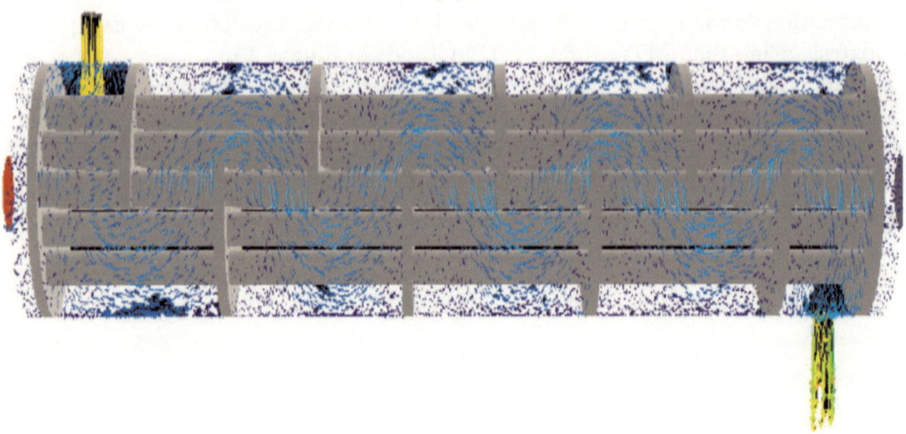

(b)

Figure 16.
The water flow pathlines on the shell side were obtained through Ansys Fluent for (a) model 1 and (b) model 2 of the heat exchanger. Case 5.

the materials involved in each section increase, obtaining a solution by solving numerically Eqs. (1)–(6) becomes complex. The system must be subdivided or meshed into several elements to get the numerical solution and avoid variation in the results. The excessive meshing and its quality imply a high computing time cost. Therefore, mesh sensitivity studies are necessary to determine the optimal analysis time. Consequently, this process can be time-consuming and impractical, unlike macroscopic analysis. **Figure 16(a)** and **(b)** show the water flow pathlines on the shell side obtained by solving the microscopic model in the geometric representation of the heat exchanger through Ansys Fluent®. Fortunately, the potential of computing systems has increased enormously in recent decades, reducing analysis times. Therefore, computational fluid dynamics (CFD) analysis is more accessible but still expensive.

3. Conclusions

Heat exchangers are cost-effective technology due to their adaptability in process optimization. This chapter analyzed use cases of exchangers, showing that with established operating parameters and correlations, the amount of heat transferred, and the fluid temperatures on the shell and tube sides are precisely quantified. Calculations in the opposite direction were also obtained using the algorithms proposed and codified in ESS. The functionality of this program for solving systems of equations facilitates the ability to create a predictive tool. Therefore, the tool applied to various operating conditions produced satisfactory results by having a database of correlations for estimating heat transfer coefficients for internal and external flows. Although the correlations are limited with the numbers of Re and Pr, the determination of the Nusselt number that allows calculating the global transfer coefficients is defined with the results of fluid temperatures on the shell and tube side. In this sense, the measurement of experimental data is required to validate the calculations and define applicable margins of the specific correlations to characterize the process of interest.

According to the above, macroscopic balances constitute an efficient methodology for the design objective of heat exchangers. However, simulations using microscopic balances represent a fundamental activity in determining a design fully adapted to a process. Even though it requires a relatively long time and expensive software resources, determining the velocity fields and temperature distribution clarifies the definition of dimensions, geometries, and materials for the construction of the exchanger structure. In this sense, fuel from non-renewable resources is used efficiently. Consequently, the carbon footprint of the process is reduced, especially if part of the thermal energy in transit continues to be used in alternate subprocesses. The above determines the concept of net zero energy to reduce costs and generate more significant benefits and is complementary to the use of renewable energies. In the case of using renewable resources as a source of thermal energy, efficiency is still an essential factor since generation is complex, and the objective is its maximum use.

Acknowledgements

This work was supported by the Universidad Politécnica de Juventino Rosas.

The authors thank the Universidad de Guanajuato for the facilities and for using the EES program to generate the graphical user interface examples for the chapter's analysis.

We also thank the Tecnológico Nacional de México, which provided access to the Ansys Fluent program through the Instituto Tecnológico de Morelia.

Conflict of interest

The authors declare no conflict of interest.

Thanks

Thanks are due to the access to the technical information provided by Dr. Juan Gregorio Hortelano Capetillo, the material discussed in this chapter was possible to concentrate on and present.

Nomenclature

ΔP_t	pressure drop in the fluid
ε	effectiveness/viscous dissipation
ε_1	effectiveness for one-step conditions
κ	turbulence kinetics
μ	viscosity
μ_t	Eddy viscosity
η	efficiency of the exchanger
ρ	density
σ_ε	constant
σ_κ	constant
τ	shear strength
A_T	total exchange area
B	spacing between deflectors
C_μ	constant
$C_{1\varepsilon}$	constant
$C_{2\varepsilon}$	constant
C_m	minor thermal capacity of the two substances in the system
C_p	thermal capacity at constant pressure
C_r	heat capacity of the fluids in the exchanger
C_s	heat capacity of shell material in the exchanger
C_t	heat capacity of tubes material in the exchanger
C_v	thermal capacity at constant volume
D_e	equivalent diameter
D_{it}	tube internal diameter
D_{ot}	tube external diameter
e	output location subscript
E	vertical distance of clearance between tangents on the surface of the tubes
E_{ij}	deformation rate components
f	friction factor estimated from the Reynolds
G_s	the mass velocity of the fluid
h_i	heat transfer coefficient associated with internal flow

h_o	heat transfer coefficient associated with external flow
H'_{fg}	latent heat modified by thermal advection effects
H_{fg}	latent heat of transformation gas liquid
i	input location subscript
k_{tm}	the conductivity of the material of the tube walls
k	thermal conductivity
L	length
L_t	tube length
L_s	shell length
$\dot{m}$	mass flow
m	exponent
n	exponent
N_b	number of deflectors
N_{s-s}	number of steps over the arrangement of tubes
N_{s-t}	number of tube passages
NTU	number of transfer units
Nu	Nusselt number
p	pressure
Pr	Prandtl number
q	heat flux
Q	transferred energy
Re_s	Reynolds number depends on the mass velocity of the fluid
Re_D	diameter-dependent Reynolds
R_{ic}	Fouling factor
t	time variable
ΔT_{ml}	logarithmic mean temperature difference
T	temperature variable
$\overline{T}$	average temperature
T_c	lowest temperature of the fluid
T_h	highest temperature of the fluid
T_{sat}	saturation temperature
T_{wall}	wall temperature
U_g	global heat transfer coefficient
v	average velocity
x	coordinate system direction
y	coordinate system direction
y^+	Y-plus is the normal distance from the solid surface's wall to the cell's centre
z	coordinate system direction
Z1	cooling of the superheated vapor zone
Z2	condensation zone
Z3	cooling below the saturation condition

Author details

Juan Gregorio Hortelano-Capetillo[1], Jorge Sergio Téllez-Martínez[2*],
José Luis Zúñiga-Cerroblanco[1], Alberto Saldaña-Robles[3], Julio César González-Juárez[2]
and Carlos Alberto González-Rodríguez[4,5]

1 Universidad Politécnica de Juventino Rosas, Santa Cruz de Juventino Rosas, GTO, México

2 Tecnológico Nacional de México/Instituto Tecnológico de Morelia, Morelia, México

3 Departamento de Ingeniería en Agricultura, DICIVA, Universidad de Guanajuato, Irapuato, GTO, México

4 División de Ingeniería Mecatrónica, Universidad Politécnica del Valle de México, Tultitlán, México

5 Facultad de Ingeniera, Universidad Nacional Autónoma de México, Ciudad Universitaria, CDMX, México

*Address all correspondence to: jorge.tm@morelia.tecnm.mx

References

[1] Kakac S, Hongtan L. Heat Exchangers: Selection, Rating and Thermal Design. Boca Raton, London, New York, Washington D.C.: CRC Press; 1998

[2] Erdogan A, Colpan CO, Cakici DM. Thermal design and analysis of a shell and tube heat exchanger integrating a geothermal based organic Rankine cycle and parabolic trough solar collectors. Renewable Energy. 2017;**109**:372-391. DOI: 10.1016/j.renene.2017.03.037

[3] Rashidi MM, Mahariq I, Alhuyi-Nazari M. Comprehensive review on exergy analysis of shell and tube heat exchangers. Journal of Thermal Analysis and Calorimetry. 2022;**147**:12301-12311. DOI: 10.1007/s10973-022-11478-2

[4] Geankoplis CJ. Transport Processes and Unit Operations. Englewood Cliffs, New Jersey: Prentice Hall International, Inc; 1993

[5] Özisik MN. Transferencia de calor – Un texto básico. Río de Janeiro: Guanabara Koogan S.A; 1990

[6] De Sandstorm of – Own job, CC BY-SA 4.0. Available from: https://commons.wikimedia.org/w/index.php?curid=133944928.

[7] Incropera FP, Dewitt DP, Bergman TL, Lavine AS. Fundamentals of Heat and Mass Transfer. 6th ed. Hoboken NJ: John Wiley & Sons; 2007

[8] Kern DQ. Process Heat Transfer. New York: McGraw Hill; 1950

[9] Crowe C, Elger D, Williams R, Roberson J. Engineering Fluid Mechanics. 9th ed. Hoboken NJ: John Wiley & Sons; 2009

[10] Potter MC, Wiggert DC, Ramadan BH. Mechanics of Fluids. 4th SI units ed. Stamford CT: Cengage Learning; 2012

[11] Coulson JM, Richardson JF. Chemical Engineering. 6th ed. Vol. 1. Jordan Hill Oxford, Woburn MA: Butterworth Heinemann; 1999

[12] Nusselt W. The condensation of steam on cooled surfaces. Zeitschrift des Vereins Deutscher Ingenieure. 1916;**60**: 541-546 and 569-575 (translated into English by Fullarton [1982] Chem. Eng. Fund. 1: 6-19)

[13] Colburn AP. A method of correlating forced convection heat transfer data and a comparison with fluid friction. International Journal of Heat and Mass Transfer. 1964;7:1359-1384

[14] Dittus FW. Boelter LMK. University of California Berkeley: Publications on Engineering; 1930. p. 2

[15] Gnielinski U. New equations for heat and mass transfer in turbulent pipe and channel flow. International Chemical Engineering. 1976;**16**:359-368

[16] Petukhov BS. Heat transfer and friction in turbulent pipe flow with variable physical properties. In: Advances in Heat Transfer. Vol. 6. New York: Academic; 1976. pp. 504-564

[17] Zukauskas AA. Convective heat transfer in cross flow. In: Kakac S, Shah RK, Aung W, editors. Handbook of Single-Phase Convective Heat Transfer. New York: Wiley; 1987. pp. 6.1-6.45

[18] Hilpert R. Fousch Geb. Ingenieurwes. 1933;**4**:215

[19] Bell KJ. Delaware method for shell side design. In: Kakac S, Bergles AE, Mayinger F, editors. Heat Exchanger – Thermal Hydraulic Fundamentals and Design. Washington D.C.: Taylor & Francis; 1981. pp. 581-618

[20] Taborek J. In: Schlunder EU, editor. Shell and Tube Heat Exchanger in Heat Exchanger Desing Handbook. New York: Hemisphere; 1983; Section 3.3

[21] Hausen H. Darstellung des Warmeuberganges in Rohrendurch verallgemeinerte Potenz beziehungen, Z. VDI Beiheft Verfahrenstechnik. 1943; **4**(1943):91-102

[22] EES. [Internet]. 2024. Available from: https://fchartsoftware.com/ees/

[23] Costa ALH, Queiroz EM. Design optimization of shell-and-tube heat exchangers. Applied Thermal Engineering. 2008;**28**(14-15):1798-1805. DOI: 10.1016/j.applthermaleng.2007.11.009

Chapter 5

Two Modeling Approaches for sCO_2-PCHEs Transients

Xin Sui and Peixin Dong

Abstract

Printed circuit heat exchangers (PCHEs) are ideal for sCO_2 power cycles due to their compactness, effectiveness, and high-pressure capability. However, their unique architecture complicates the modeling of their dynamic behavior in power cycles, which experience rapid transients. Both the significant computational resources required and the high investment costs of experiments limit their widespread application. To model the component- and system-level transients of sCO_2-PCHEs, this study presents two 1D modeling approaches for different purposes: one for component-level simulation based on local properties of sCO_2, and the other for system-level simulation based on transfer functions. Given the significant discrepancies observed when using a fixed time constant in the latter approach, this paper introduces the concept of an optimal time constant to model the transient behavior of PCHEs as a first-order system with minimal prediction error. This optimal time constant varies with operating conditions, contrary to what the name might suggest. These results demonstrate the potential of PCHEs in advanced power cycles and provide valuable insights for accurate system-level sCO_2 power cycle control studies.

Keywords: printed circuit heat exchangers (PCHEs), transient behaviors, 1D numerical model, low-order approximation, optimal time constant

1. Introduction

Supercritical carbon dioxide (sCO_2) power cycles are promising solutions for transforming traditional thermal powers to renewable energies [1, 2]. The sCO_2 cycles find massive applications across vast working scenarios thanks to their high conversation efficiency but low carbon emissions. Among various types of heat exchangers, printed circuit heat exchangers (PCHEs) are favored as primary heaters and recuperators in sCO_2 cycles because of their outstanding heat transfer performance and remarkable resistance to high temperature and pressure. Massive experimental and numerical studies have been conducted to investigate sCO_2-PCHEs' dynamic performance under varying scenarios, such as nuclear power [3, 4], concentrated solar power [1], wasted heat recovery [5, 6] and hypersonic aero-engines [7]. For these investigations, the most challenging task is to handle the difficulties brought by sCO_2 fluid's steep property variations.

However, the temperature-dependent properties violate the assumption of 'constant property' employed by most dynamic modeling approaches [8, 9], resulting in significant and lagging prediction errors, for which both three- and one-dimensional simulation techniques have paid special efforts to address the issue. The high-fidelity three-dimensional modeling approaches, such as Computational Fluid Dynamics (CFD) [10, 11] and Direct Numerical Simulation (DNS), capture the dramatic property variations at the price of magnificent computation cost [12]. The one-dimensional modeling methods prefer the discretized approach to capture local properties while requiring remarkable computational costs [13]. On the other hand, AI-assisted technologies have shown promise in enhancing power plant efficiency, but their application in predicting system-level off-design performance and optimizing control strategies remains underexplored [14, 15]. This limitation is primarily due to the scarcity of operational data needed to train AI models effectively. The data, which are derived from experimental tests, require significant investment costs, making them accessible only to a few well-equipped laboratories [16].

Utilizing inappropriate empirical correlations might induce unexpected convergence and stability issues to dynamic modeling [17, 18]. The dynamic simulation, aiming to provide insightful knowledge of sCO_2-PCHEs' transients for control studies, prefers the lower-order system to proxy PCHEs' transients with acceptable errors. While experimental observations and analytical solutions have confirmed the feasibility of using the first-order system approximation for shell-and-tube and plate heat exchangers [19, 20], the same modeling technique has not been reported for sCO_2-PCHEs, as we know. Therefore, this chapter will propose a component-level 1D model based on sCO_2 local properties and a first-order approximation for sCO_2-PCHEs. The feasibility of both models under different operating conditions are compared. A data-fitting approach will also be presented to compute the key parameter, time constant, of the proposed low-order approximation. As will be seen, a varying time constant, determined by both geometrical parameters and operating conditions, can significantly reduce the approximation's errors. By the end of this study, the transients of sCO_2 PCHEs can be approximated with minimal cost both experimentally and numerically, with good accuracy.

2. Modeling methodologies

2.1 1D numerical model for sCO_2-PCHEs

A sCO_2 Recompression Brayton Cycle (**Figure 1a**) usually chooses PCHEs as its precooler, high-temperature and low-temperature recuperator. The sCO_2 fluid experiences dramatic thermal-hydraulic variations in these components, as shown in the changes of Prandtl numbers in **Figure 1b** and **c**.

To investigate the dynamics of sCO_2 power cycles, 1-D numerical modeling [2] usually discretizes a straight- or zigzag-channel PCHE into $N-1$ heat transfer units (**Figure 2**), where each unit consists of one hot volume, one cold volume, and one separating metal wall volume. For the j_{th} unit $j=(1,\dots,N-1)$, the governing conservation equations for mass, energy, and momentum for either the cold or hot side are,

$$\frac{\partial M_j}{\partial t} = A\cdot l\cdot\left[\frac{\partial \rho_{:j}}{\partial p_{:j}}\cdot\dot{p}_{:j} + \frac{\partial \rho_{:j}}{\partial T_{:j}}\cdot\dot{T}_{:j} + \frac{\partial \rho_{:j+1}}{\partial T_{:j+1}}\cdot\dot{T}_{:j+1}\right] + \dot{w}_{:j} - \dot{w}_{:j+1} \qquad (1)$$

$$A \cdot l \cdot \rho_j \cdot c_{v,j} \cdot \dot{T}_j = w_{:j+1} \cdot h_{:j+1} - w_{:j} \cdot h_{:j} + Q_{x,j} \tag{2}$$

$$\frac{l}{A}\frac{\partial w_j}{\partial t} = p_{:j} - p_{:j+1} - \Delta p_{fric,j} \tag{3}$$

where the mass M_j, averaged density ρ_j, averaged pressure p_j, averaged temperature T_j, averaged specific heat capacity $c_{v,j}$, and mass flow rate w_j are volume-average

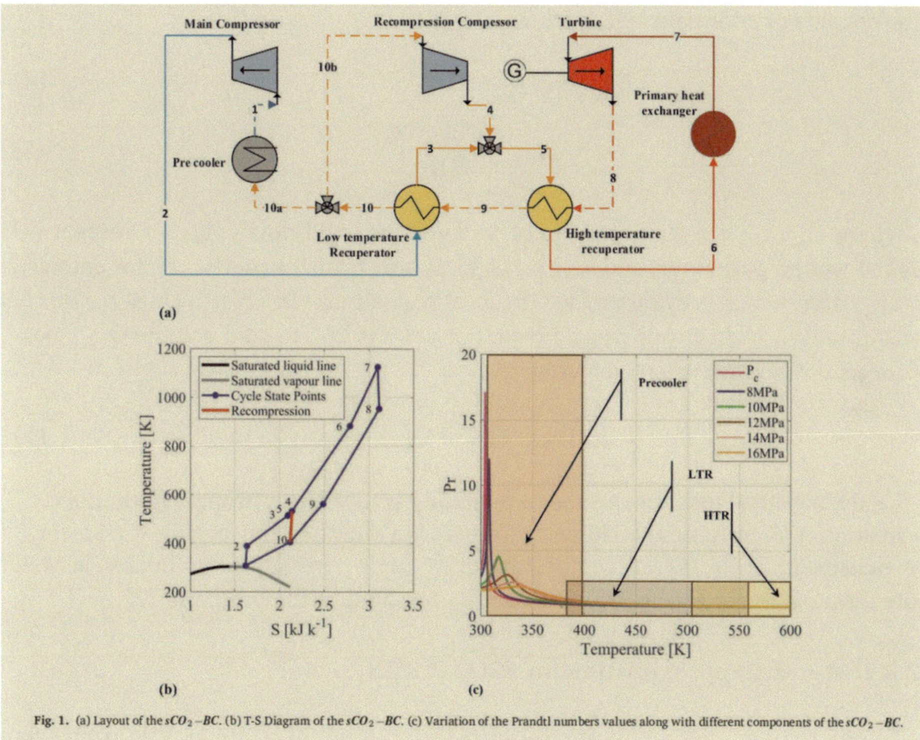

Figure 1.
sCO_2's property variations per components of recompression Brayton cycles.

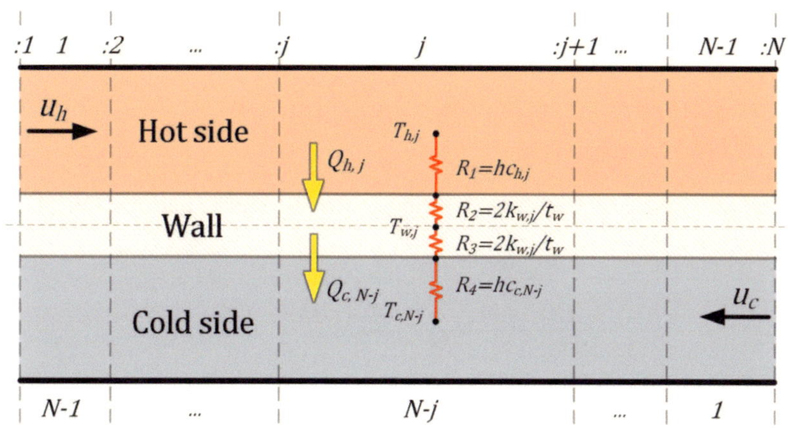

Figure 2.
Schematic of discretized PCHE model.

quantities of volume j, A is the cross-sectional area (identical for node and volume), $l = L/(N-1)$ is the volume length, $\rho_{:j}, p_{:j}, T_{:j}, h_{:j}, w_{:j+1}$ are averages of density, pressure, temperature, specific enthalpy, and mass flow rate over the cross-section area of node : j, and $\Delta p_{fric,j}$ is the local pressure drop through the volume caused by the friction.

For the energy conservation equation (Eq. (2)), $Q_{x,j}$ denotes the heat transferred between a flow volume j and its adjacent wall volume j, where the subscript x is either h for hot side or c for cold side, such as,

$$Q_{x,j} = \left(T_{w,j} - T_{x,j}\right) \cdot \omega \cdot l \cdot hc_{x,j} \tag{4}$$

$$hc_{x,j} = \frac{\mathrm{Nu}_{x,j} \cdot k_{x,j}}{D_{hyd}} \tag{5}$$

where $T_{w,j}, T_{x,j}$ are temperatures of wall volume j and flow volume j, respectively, ω is the wetted perimeter, and $hc_{x,j}$ is the local heat transfer coefficient for convective heat transfer, which is evaluated via the local Nusselt number $\mathrm{Nu}_{x,j}$. Thus, for this heat transfer unit j, the heat transferred from hot volume $Q_{h,j}$ to cold volume $Q_{c,N-j}$ is balanced at the in-between wall volume,

$$l \cdot \overline{\rho}_w \cdot A_w \cdot \dot{T}_{w,j} = Q_{h,j} + Q_{c,N-j} \tag{6}$$

Lastly, a critical task for the above modeling is to select a proper Nu number correlation to determine the Nusselt numbers Nu and then the local heat transfer coefficients $hc_{x,j}$ (Eq. (5)). A comparison of multiple correlation candidates, as well as more discussions for the above models, are presented in the literature [2].

2.2 A first-order approximation for sCO$_2$-PCHEs

As reported in several published studies, the exit temperature of heat exchangers can be expressed as a first-order approximation [19, 20]. In specific, if one inlet is experiencing a step change in mass flow rate or temperature, the exit temperature is a function of time constant τ and time delay t_r, such as,

$$\begin{cases} t \le t_r: & T(t) = T_0 \\ t > t_r: & \dfrac{T(t) - T_\infty}{T_0 - T_\infty} = \exp\left(-\dfrac{t - t_r}{\tau}\right) \end{cases} \tag{7}$$

Using this approximation, the dynamic behavior of the heat exchanger can be represented as a first-order system described by a transfer function (TF) [17, 21], such as,

$$G(s)_{1st} = \frac{Y(s)}{F(s)} = \frac{K}{\tau s + 1} \exp(-t_r s) \tag{8}$$

where $Y(s), F(s)$ are the Laplace transformation of system's response $y(t)$ and forcing input $f(t)$, respectively, τ, time delay t_r and gain K are to-be-determined parameters. For sCO$_2$-PCHEs, the time delay t_r usually equals 0 regarding PCHEs' short flow path, and the gain K normally equals 1 for the symmetry of two flows, i.e., a step change in one flow would lead to a change of the same scale in the other flow. Therefore, this chapter will demonstrate how to calculate the time constant τ with great attention.

2.3 Time constant calculation

The time-domain response of the first-order system (Eq. (8)), when taking a step-down input $f(t) = Au(t)$, can be described as,

$$T(t) = T_0 e^{-t/\tau} + KAu(t - t_r)\left(1 - e^{-(t-t_r)/\tau}\right) \tag{9}$$

where T_0 is the initial value of $T(t)$ for $t \leq t_r$. Therefore, we can locate the best guess of time constant through the data-fitting approach. The retrieved time constant τ as well as gain K and delayed time t_r should fit the system with the experimental or simulation results with minimum errors [22, 23], such as,

$$\underset{\tau, K, t_r \in R}{\arg\min} \frac{1}{n} \sum_{i=1}^{n} \left[T_i - y_i\right]^2 \tag{10}$$

where $T_i = T(t = t_i)$ is the time-domain response of the system (Eq. (9)) at time t_i, y_i is the PCHE's outlet temperature at the same moment, $i = 1, \ldots, n$ indexes n moments of a dynamic experiment or simulation process. In the above optimization, mean square errors (*MSE*s) are employed to evaluate prediction errors, for which maximum absolute error σ_{max} is an alternative metric, such as,

$$\sigma_{max} = \max\left(\left|T_i - y_i\right|\right), \quad i = 1, \ldots, n \tag{11}$$

3. Validation case

3.1 Set-ups

This section sets up a parameterized, general-purpose Modelica program to simulate sCO$_2$ PCHEs' dynamic behavior under various experimental scenarios, with which we can calculate time constant τ accordingly. The program, as published in our open source project (SCOPE) [2], adopts two source mass flow components and two sink pressure components. By setting the temperature, mass flow rate, and pressure of the sources to the experimental scenarios, the program can simulate the PCHE's dynamic behavior under various geometrical parameters and operating conditions. The graphical user interface of the simulation program is illustrated in **Figure 3**.

3.2 First-order system validation

We select the configuration (**Table 1**) of a HTR of a 10 WMe sCO$_2$ Recompressed Brayton Cycle [24] as a demonstration. When taking a step-down change in hot inlet mass flow rate, the hot outlet temperature exponentially drops as defined in Eq. (9). The time constant τ identified by Eq. (10) is 251.92 s, which yields a good match between SCOPE's simulation output and the first-order system, with an MSE of $0.6761^{\circ}C^2$ and a $\sigma_{max} = 1.4188^{\circ}C$ (**Figure 4a, Table 2**).

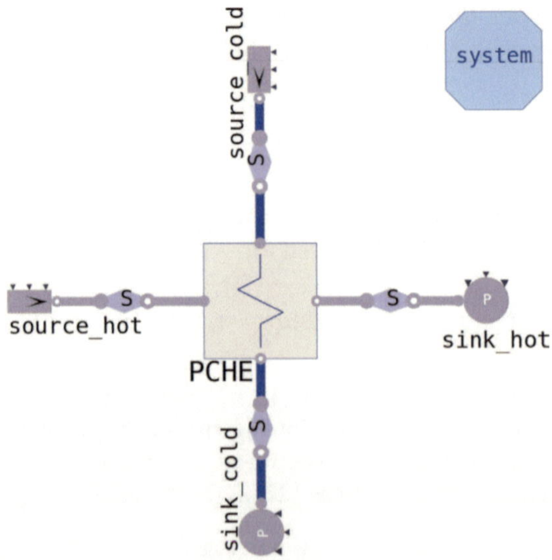

Figure 3.
Modelica test program GUI.

Name	Symbol	Value
Hot/Cold inlet temperature	$T_{h,in}/T_{c,in}$ (°C)	578.15 / 181.95
Hot/Cold outlet temperature	$T_{h,out}/T_{c,out}$ (°C)	192.0 / 532.1
Hot/Cold inlet pressure	$P_{h,in}/P_{c,in}$ (bar)	89.62 / 237.48
Hot/Cold mass flow rate	$\dot{m}_h/\dot{m}_c$ (kg/s)	100.17 / 100.17
Weight / Height	W/H (mm)	360 / 3710
Channel length	L_{fp} (mm)	800
Channel diameter	D_{ch} (mm)	1.5
Metal wall thickness	T_{wall} (mm)	0.51
Channel number	N_{ch} ([1])	700,000

Table 1.
Geometry and operation parameters for time constant evaluation tests.

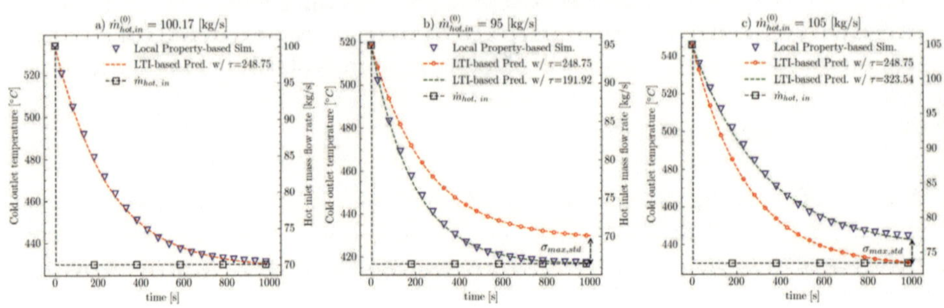

Figure 4.
Comparison between local property-based simulation and LTI system prediction.

Name case	Standard	Off-Des. case A	Off-Des. case B
$\dot{m}_{hot,in}$	100.17 kg/s	95 kg/s	105 kg/s
τ_{op}	251.92	197.66	320.08
MSE	0.6761	0.3209	1.2938
σ_{max}	1.4189	2.1662	1.8018
MSE_{std}	–	206.7400	200.5752
$\sigma_{max,std}$	–	17.2889	17.3021

Table 2.
Detailed discrepancies between standard case and two off-design cases.

However, the same value of τ will produce significant errors when applying for other starting mass flow rates (**Figure 4b** and **c**). For a below-design-point operation where $\dot{m}_{hot,in}$ starts from 95 kg/s (**Figure 4b**), the hot outlet temperature predicted by the first-order system, with $\tau = 251.92$, diverges from the SCOPE's result by an MSE of $206.74^{\circ}C^2$ and an $\sigma_{max,\ std}$ of $17.2889°C$ at $t = 1000\ s$. Similarly, for an above-design-point operation where $\dot{m}_{hot,in}$ starts from 105 kg/s, the discrepancies are $MSE = 320.08^{\circ}C^2$ and $\sigma_{max,\ std} = 17.3021^{\circ}C$ at $t = 1000\ s$. The remarkable discrepancies reflect the drawback of the assumption of constant heat transfer coefficients held by the low-order system approximation. Under the assumption, the time constant is a system instinct property determined by heat exchangers' geometrical parameters and should remain constant during a system's transient. However, for sCO_2-PCHEs where the working fluid experiences steep non-linear variations, the time constant's value varies as the initial operating condition changes. As a consequence, employing the τ identified at the on-design condition will produce significant prediction discrepancies if the PCHE works at alternative operating conditions (**Figure 4b** and **c**). A variable time constant, on the contrary, can achieve minimal prediction errors under various operating conditions.

To prove the above point, we identify the value of time constants per operating condition, which produces good agreements between the first-order system and SCOPE's outputs. In specific, a time constant τ valued at 197.66 produces a prediction with $MSE = 0.3209^{\circ}C^2$ and $\sigma_{max} = 2.1662^{\circ}C$ for the below-design-point operation (**Figure 4b**) and a time constant τ valued at 320.08 generates an $MSE = 1.2938^{\circ}C^2$ and $\sigma_{max} = 1.8018^{\circ}C$ for the above-design-point case (**Figure 4c**). Therefore, it is reasonable to introduce a notation named optimal time constant, τ_{op}, as a refined key parameter to generate minimal prediction errors for a sCO_2 PCHE applications under varying configurations toward dynamic and control strategic studies.

4. Conclusions

For sCO_2-PCHEs, a variable time constant determined by the system's operating conditions is favorable to approximating the component's dynamic behavior with a first-order system. Moreover, PCHEs are critical components for sCO_2 power blocks considering their substantial impacts on the heat transformation and flexibility of the entire power block. A better quantitative understanding of its flexibility through analyzing time constant can effectively serve the investigation of the control strategy of sCO_2 power cycles. Understanding the τ_{op}'s distribution under varying operating conditions provides informative knowledge for designing and optimizing control strategies. For example, it is reasonable to maintain the working fluids at relatively low initial mass flow rates and alter the mass flow rate with the highest possible magnitude (within the permitted limit) to achieve the lowest time constant and highest flexibility. Therefore, for power block control studies, it is essential to take the trade-off between τ and η into account and utilize the advantages to achieve the control objective with the highest speed but lowest cost.

Author details

Xin Sui[1,2] and Peixin Dong[3*]

1 School of Mechatronics Engineering, Henan University of Science and Technology, Luoyang, Henan, China

2 Key Laboratory of Mechanical Design and Transmission System of Henan Province, Luoyang, Henan, China

3 Department of Mechanical Engineering, The Hong Kong Polytechnic University, Hong Kong SAR, China

*Address all correspondence to: p.dong@uqconnect.edu.au

References

[1] Lock A, Bone V. Off-design operation of the dry-cooled supercritical CO_2 power cycle. Energy Conversion and Management. 2022;**251**:114903. DOI: 10.1016/j.enconman.2021.114903

[2] Sui X, Russell H, Gurgenci H, Lee S, Guan Z, Dong P. A one-dimensional modelling methodology of printed circuit heat exchangers for steady, off-design and transient feasibility in a supercritical CO_2 power block. International Journal of Heat and Mass Transfer. 2022;**193**:122938. DOI: 10.1016/j.ijheatmasstransfer.2022.122938

[3] Park JH, Kwon JG, Kim MH, Cha JE, Jo H. Experimental investigation of buoyancy effects on local heat transfer of supercritical pressure CO_2 in horizontal semicircular tube. International Journal of Heat and Mass Transfer. 2021;**164**:120496. DOI: 10.1016/j.ijheatmasstransfer.2020.120496

[4] Park JH, Kwon JG, Kim TH, Kim MH, Cha J-E, Jo H. Experimental study of a straight channel printed circuit heat exchanger on supercritical CO_2 near the critical point with water cooling. International Journal of Heat and Mass Transfer. 2020;**150**:119364. DOI: 10.1016/j.ijheatmasstransfer.2020.119364

[5] Zhang H, Shi L, Xuan W, Chen T, Li Y, Tian H, et al. Analysis of printed circuit heat exchanger (PCHE) potential in exhaust waste heat recovery. Applied Thermal Engineering. 2022;**204**:117863. DOI: 10.1016/j.applthermaleng.2021.117863

[6] Alfani D, Binotti M, Macchi E, Silva P, Astolfi M. sCO_2 power plants for waste heat recovery: Design optimization and part-load operation strategies. Applied Thermal Engineering. 2021;**195**:117013. DOI: 10.1016/j.applthermaleng.2021.117013

[7] Li H, Liu H, Zou Z. Experimental study and performance analysis of high-performance micro-channel heat exchanger for hypersonic precooled aero-engine. Applied Thermal Engineering. 2021;**182**:116108. DOI: 10.1016/j.applthermaleng.2020.116108

[8] Liu G, Huang Y, Wang J, Liu R. A review on the thermal-hydraulic performance and optimization of printed circuit heat exchangers for supercritical CO_2 in advanced nuclear power systems. Renewable and Sustainable Energy Reviews. 2020;**133**:110290. DOI: 10.1016/j.rser.2020.110290

[9] Bone V, McNaughton R, Kearney M, Jahn I. Methodology to develop off-design models of heat exchangers with non-ideal fluids. Applied Thermal Engineering. 2018;**145**:716-734. DOI: 10.1016/j.applthermaleng.2018.09.082

[10] Saeed M, Berrouk AS, Salman Siddiqui M, Ali Awais A. Numerical investigation of thermal and hydraulic characteristics of sCO_2-water printed circuit heat exchangers with zigzag channels. Energy Conversion and Management. 2020;**224**:113375. DOI: 10.1016/j.enconman.2020.113375

[11] Saeed M, Ali Awais A, Berrouk AS. CFD aided design and analysis of a precooler with zigzag channels for supercritical CO_2 power cycle. Energy Conversion and Management. 2021;**236**:114029. DOI: 10.1016/j.enconman.2021.114029

[12] Zhou L, Garg D, Qiu Y, Kim S-M, Mudawar I, Kharangate CR. Machine learning algorithms to predict flow condensation heat transfer coefficient in mini/micro-channel utilizing universal data. International Journal of Heat and Mass Transfer. 2020;**162**:120351. DOI: 10.1016/j.ijheatmasstransfer.2020.120351

[13] Kwon JS, Bae SJ, Heo JY, Lee JI. Development of accelerated PCHE off-design performance model for optimizing power system operation strategies in s-CO_2 Brayton cycle. Applied Thermal Engineering. 2019;**159**: 113845. DOI: 10.1016/j.applthermaleng.2019.113845

[14] Shahzad MW, Nguyen VH, Xu BB, Tariq R, Imran M, Ashraf WM, et al. Machine learning assisted prediction of solar to liquid fuel production: A case study. Process Safety and Environmental Protection. 2024;**184**:1119-1130. DOI: 10.1016/j.psep.2024.02.060

[15] Muhammad Ashraf W, Moeen Uddin G, Afroze Ahmad H, Ahmad Jamil M, Tariq R, Wakil Shahzad M, et al. Artificial intelligence enabled efficient power generation and emissions reduction underpinning net-zero goal from the coal-based power plants. Energy Conversion and Management. 2022;**268**:116025. DOI: 10.1016/j.enconman.2022.116025

[16] Yang J, Yang Z, Duan Y. A review on integrated design and off-design operation of solar power tower system with s-CO_2 Brayton cycle. Energy. 2022; **246**:123348

[17] Chen J, Cui G, Xiao Y. An analytical solution to the dynamic behavior of heat exchanger networks. International Journal of Heat and Mass Transfer. 2018; **126**:466-478. DOI: 10.1016/j.ijheatmasstransfer.2018.05.041

[18] Yang R, Tran CT. An analytical heat exchanger model to study dynamic behavior in the case of simultaneous inlet variations. International Journal of Thermal Sciences. 2022;**174**:107452. DOI: 10.1016/j.ijthermalsci.2021.107452

[19] Lachi M, El Wakil N, Padet J. The time constant of double pipe and one pass shell-and-tube heat exchangers in the case of varying fluid flow rates. International Journal of Heat and Mass Transfer. 1997;**40**(9):2067-2079. DOI: 10.1016/S0017-9310(96)00274-8. Available from: https://www.sciencedirect.com/science/article/pii/S0017931096002748

[20] Das SK, Spang B, Roetzel W. Dynamic behavior of plate heat exchangers—Experiments and modeling. Journal of Heat Transfer. 1995;**117**(4):859-864. DOI: 10.1115/1.2836302

[21] Cui GM, Wang JY, Jiang T, Hu XB. Transfer behavior of heat exchanger. Advanced Materials Research. 2011; **233–235**:1281-1287. DOI: 10.4028/www.scientific.net/AMR.233-235.1281

[22] Alfani D, Astolfi M, Binotti M, Campanari S, Casella F, Silva P. Multi objective optimization of flexible supercritical CO_2 coal-fired power plants. In: Volume 3: Coal, Biomass, Hydrogen, and Alternative Fuels; Cycle Innovations; Electric Power; Industrial and Cogeneration; Organic Rankine Cycle Power Systems. Phoenix, Arizona, USA: American Society of Mechanical Engineers; 2019. p. V003T06A027. DOI: 10.1115/GT2019-91789

[23] Ma T, Li M-J, Xu J-L, Ni J-W, Tao W-Q, Wang L. Study of dynamic response characteristics of s-CO_2 cycle in coal-fired power plants based on real-time micro-grid load and a novel synergistic control method with variable

working conditions. Energy Conversion and Management. 2022;**254**:115264. DOI: 10.1016/j.enconman.2022.115264

[24] Jiang Y, Liese E, Zitney SE, Bhattacharyya D. Design and dynamic modeling of printed circuit heat exchangers for supercritical carbon dioxide Brayton power cycles. Applied Energy. 2018;**231**:1019-1032. DOI: 10.1016/j.apenergy.2018.09.193

Section 3

Material Thermodynamics

Chapter 6

Dynamic Specific Heat and Glass Transitions

Seiji Kojima

Abstract

The dynamical properties such as fragility, and non-Debye behavior of glass-forming materials have been studied by the frequency-dependent dynamic specific heat. Kubo's formula on the fluctuation-dissipation theorem defines dynamic specific heat using the correlation function of enthalpy fluctuations. Dynamic specific heat is important for analyzing and understanding various relaxation processes. The dielectric relaxation is caused only by polar atomic motions, while the enthalpy relaxation is caused by total degrees of freedom of atomic motions. This chapter introduces two experimental methods to measure dynamic specific heat: (1) temperature-modulated differential scanning calorimetry (MDSC) and (2) photoacoustic spectroscopy. The experimental results of the dynamical properties of glass transitions in oxide glasses with covalent bond network structures and hydrogen-bonded glass-forming materials are reviewed.

Keywords: dynamic specific heat, modulated DSC, photoacoustic spectroscopy, relaxation, fragility, glass transition

1. Introduction

When a simple liquid is cooled from a high temperature, it undergoes a liquid-solid phase transition and crystallizes below a melting temperature, T_m. In contrast, when a complex liquid is cooled down below T_m, it changes into a supercooled liquid, which is a nonequilibrium state. Upon further cooling, a liquid-glass transition occurs at a glass transition temperature, T_g and it changes into a nonequilibrium glassy state. **Figure 1** shows the enthalpy of liquid phase (AB), supercooled liquid state (BD), glass state (DE), and crystal phase (CG) as a function of temperature. The enthalpy of a liquid phase and supercooled liquid state crosses to that of a crystal phase at the Kauzmann temperature [1], T_K (F), which is an ideal glass transition temperature determined by the static condition. Another ideal glass transition temperature is the Vogel-Fulcher temperature [2], T_{VF}, which is determined by the Vogel-Fulcher law on the main structure relaxation process. In most glasses T_K is close to T_{VF}.

Upon cooling a supercooled liquid, the relaxation time of a main structural relaxation process increases and undergoes a liquid-glass transition into a glass state at, T_g, where the relaxation time is 100 s. Both supercooled liquid and glass belong to nonequilibrium states. Understanding the dynamical properties of glass-forming

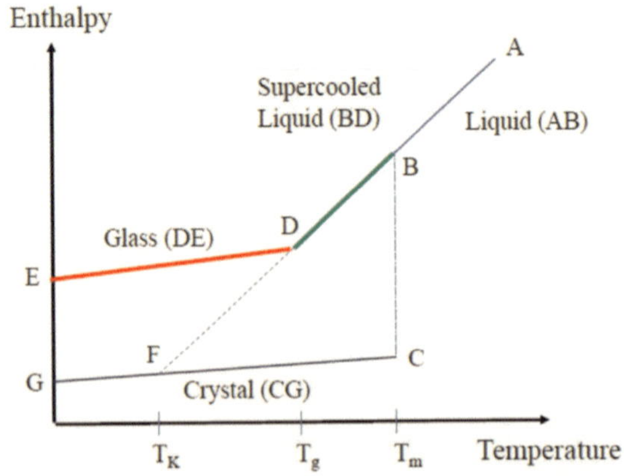

Figure 1.
The enthalpy is plotted as a function of temperature. Above T_m only a liquid phase (AB) exists. Below T_m a simple liquid crystallizes and crystal phase (CG) appears. While, a complex liquid keeps a supercooled liquid state (BD) between T_m and T_g. Below T_g a glass state (DE) appears.

materials is very important to clarify the mechanism of a liquid-glass transition. Their structural relaxation processes have been extensively studied by dielectric and ultrasonic spectroscopies, which are responsible for dielectric and elastic relaxations, respectively [3, 4].

According to the linear-response theory, dynamic specific heat is defined by Kubo's formula on the fluctuation-dissipation theorem using the correlation function of enthalpy fluctuations [5]. The dynamic-specific heat provides a general understanding on the relaxation phenomena. Because the enthalpy relaxation contains the total degrees of freedom of atomic motions including polar atomic motions. In contrast, dielectric relaxation contains only polar atomic motions, and other nonpolar atomic motions do not contribute to the dielectric relaxation [6]. Therefore, it is important to study the structural relaxation processes of glass-forming materials by the frequency-dependent dynamic specific heat [7].

2. Dynamic susceptibility

Various physical quantities show time-dependent responses when an external field is applied to a system to be studied. Such a time-dependent response in a nonequilibrium state is analyzed by a response function or a relaxation function in linear-response theory, assuming that an external field is too weak to induce a nonlinear response. In observing a time-dependent response of physical quantities in a frequency-domain, dynamical susceptibility shows a frequency dispersion. The frequency dispersions of complex dielectric constants and complex elastic constants have been measured to study dynamical processes in a nonequilibrium state.

2.1 Complex dielectric susceptibility

In nonequilibrium statistical physics, dynamic (or complex) dielectric susceptibility is defined by the fluctuation-dissipation theory on the fluctuation of electric flux density, $D(t)$ [5]. In the classical limit, it holds that

$$\varepsilon(\omega) = \varepsilon_0 + \frac{i\omega}{k_B T}\int_0^{\infty}\langle [D(t) - \overline{D}][D(0) - \overline{D}]\rangle e^{i\omega t}dt. \quad (1)$$

Here, k_B is Boltzmann constant, $D(t)$ is electric flux density as a function of time, $\langle\rangle$ is an expected value, and $\overline{D}$ is the average value of $D(t)$.

Dynamic/complex dielectric susceptibility has been discussed by linear-response theory. When the response of electric flux density, $D(t)$, is linearly proportional to the external field, $E(t)$, dielectric flux density is given by,

$$D(t) = \varepsilon_\infty E(t) + (\varepsilon_0 - \varepsilon_\infty)\int_{-\infty}^{t}\phi(t - t')E(t')dt'. \quad (2)$$

Here ε_∞, ε_0, and $\phi(t)$ are dielectric constant of high-frequency limit, static dielectric constant, and dielectric response function, respectively. **Figure 2** shows the impulse response of $D(t)$.

By Fourier transformation from time-domain to frequency-domain, it holds that

$$D(\omega) = \int_0^{\infty} D(t)e^{i\omega t}dt = \varepsilon(\omega)E(\omega). \quad (3)$$

Here, the complex dielectric constant, $\varepsilon(\omega)$, is related to a dielectric response function, $\phi(t)$.

$$\varepsilon(\omega) = \varepsilon_\infty + (\varepsilon_0 - \varepsilon_\infty)\int_0^{\infty}\phi(t)e^{i\omega t}dt, \quad (4)$$

Dielectric relaxation function, $\Psi(t)$, is defined by

$$\phi(t) = -\frac{d}{dt}\Psi(t). \quad (5)$$

By the substitution of Eq. (5) for Eq. (4), we obtain

$$\varepsilon(\omega) = \varepsilon_0 + (\varepsilon_0 - \varepsilon_\infty)i\omega\int_0^{\infty}\Psi(t)e^{i\omega t}dt \quad (6)$$

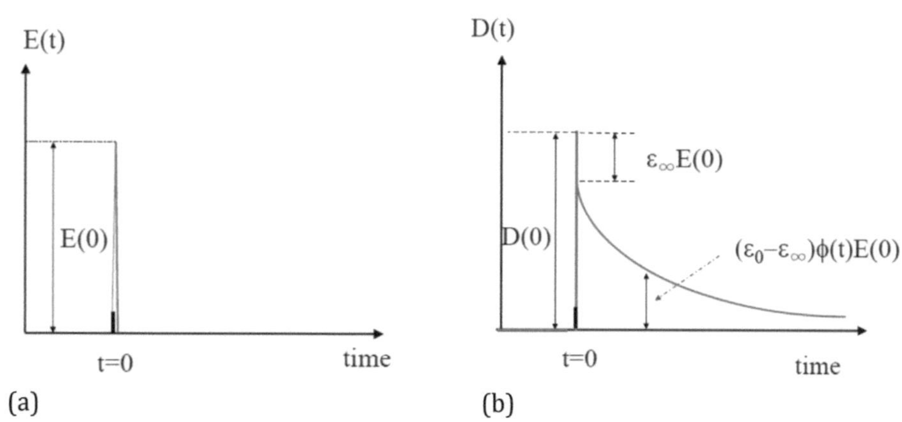

Figure 2.
(a) Application of impulse electric field with a peak height E(0) *at* t = 0. *(b) Impulse response of* D(t), *where* ϕ(t) *is a response function and* D(0) = ε_0E(0).

In the Debye model with a relaxation time τ [8], the relaxation function is given by

$$\Psi(t) = exp\left(-\frac{t}{\tau}\right). \tag{7}$$

The complex dielectric constant is given by

$$\varepsilon(\omega) = \varepsilon_\infty + \frac{\{\varepsilon_0 - \varepsilon_\infty\}}{1 + i\omega\tau}. \tag{8}$$

Real and imaginary parts of complex dielectric constants are given by

$$\varepsilon'(\omega) = \varepsilon_\infty + \frac{\varepsilon_0 - \varepsilon_\infty}{1 + \omega^2\tau^2}, \varepsilon''(\omega) = \frac{(\varepsilon_0 - \varepsilon_\infty)\omega\tau}{1 + \omega^2\tau^2}. \tag{9}$$

Figure 3 shows real and imaginary parts of complex dielectric constant with ε_∞ = 20, ε_0 = 50, and $\tau = 1.6 \times 10^{-9}$ s.

As the frequency increases from a low frequency, the real part gradually decreases from the low-frequency limit, ε_0, and in the vicinity of $\omega = 1/\tau$ remarkable decreases. For further increase in frequency, it approaches the high-frequency limit, ε_∞. The imaginary part shows the maximum at $\omega = 1/\tau$ and $\varepsilon = (\varepsilon_0 - \varepsilon_\infty)/2$. In a liquid-glass transition, upon cooling from a high-temperature liquid phase, the relaxation time of structural relaxation becomes longer toward a liquid-glass transition temperature, T_g. In a dielectric measurement, the relaxation time is determined by the maximum frequency of the imaginary part of the dielectric constant by the equation $\omega\tau = 1$.

Figure 4 shows the temperature dependence of the observed imaginary parts of glass-forming propylene glycol with T_g = 173 K in the large frequency range from 1 mHz to 10 GHz [9]. The broadband dielectric spectra were measured by the combination of two kinds of measurement. The low-frequency range between 1 mHz and 10 MHz was measured by the Impedance/Gain-phase analyzer (Solartron SI 1260). The high-frequency range between 1 MHz and 10 GHz was measured by the time-domain reflectometry (TDR) system, which is the combination of HP54750A digital

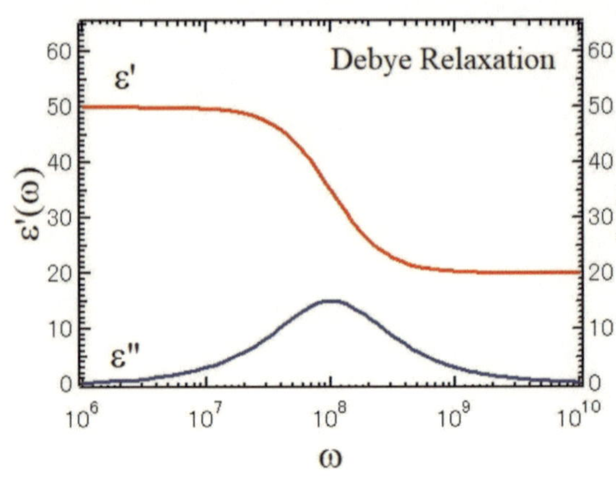

Figure 3.
Real and imaginary parts of complex dielectric constant with ε_∞ = 20, ε_0 = 50, *and* $\tau = 1.6 \times 10^{-9}$ *s.*

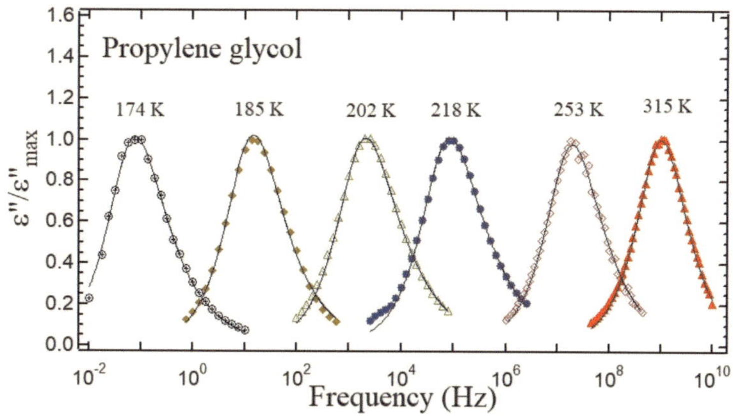

Figure 4.
The temperature and frequency dependences of the imaginary part of dielectric susceptibility of glass-forming propylene glycol.

Oscilloscope and HP54754A Differential TDR module. Upon cooling from a high temperature, the peak frequency at 315 K in a GHz range remarkably decreases down to that in a mHz range. The structural relaxation time obeys the Vogel-Fulcher-Tammann law.

In the Debye relaxation model, the equation of semi-circle holds for $\varepsilon'(\omega)$and $\varepsilon''(\omega)$.

$$\left\{\varepsilon'(\omega) - \frac{\varepsilon_0 + \varepsilon_\infty}{2}\right\}^2 + \{\varepsilon''(\omega)\}^2 = \left(\frac{\varepsilon_0 - \varepsilon_\infty}{2}\right)^2 \tag{10}$$

The plot of $\varepsilon'(\omega)$ versus $\varepsilon''(\omega)$ is called a Cole-Cole plot as shown in **Figure 5**.

In the non-Deby case, the relaxation time has a distribution, and a complex dielectric constant is given by the summation of relaxation processes with a distribution function of relaxation time, $g(\tau)$:

$$\varepsilon(\omega) = \varepsilon_\infty + (\varepsilon_0 - \varepsilon_\infty)\int_0^\infty \frac{g(\tau)}{1 + i\omega\tau} d\tau. \tag{11}$$

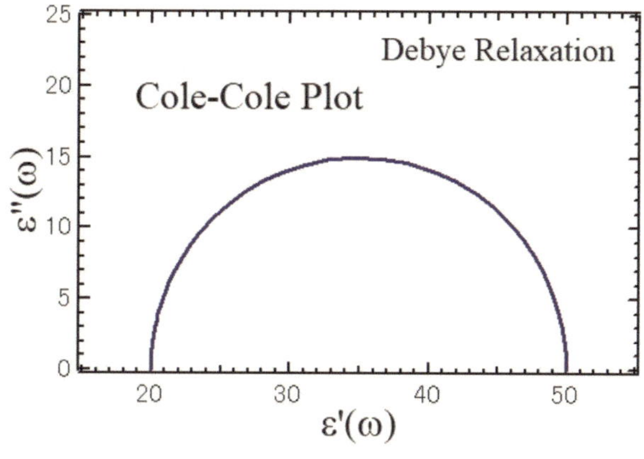

Figure 5.
Cole-Cole plot of a complex dielectric constant.

If the complex dielectric constant over a very large frequency is available, the distribution function can be determined by inverse-integral transformation. However, for the analysis of experimental results with a limited frequency range, the following empirical formulae with few fitting parameters have been used.

1. Davidson-Cole equation [10]

$$\varepsilon(\omega) = \varepsilon_\infty + \frac{\varepsilon_0 - \varepsilon_\infty}{(1 - i\omega\tau)^\gamma}, 0 < \gamma < 1 \tag{12}$$

2. Cole-Cole equation [11]

$$\varepsilon(\omega) = \varepsilon_\infty + \frac{\varepsilon_0 - \varepsilon_\infty}{1 - (i\omega\tau)^\alpha}, 0 < \alpha < 1 \tag{13}$$

3. Haviliak-Negami equation [12]

$$\varepsilon(\omega) = \varepsilon_\infty + \frac{\varepsilon_0 - \varepsilon_\infty}{\{1 - (i\omega\tau)^\alpha\}^\lambda}, 0 < \alpha, \lambda < 1 \tag{14}$$

In the time-domain, Kohlrausch-Wiliams-Watts (KWW) function has been used in non-Debye relaxation [13, 14]:

$$\Psi(t) = exp\left\{-\left(\frac{t}{\tau}\right)^\beta\right\}, 0 \le \beta \le 1 \tag{15}$$

The numerical study on the relation of parameters between time-domain KWW function and the frequency-domain Haviliak-Negami equation was reported [15].

$$ln\left(\frac{\tau_{HN}}{\tau_{KWW}}\right) \cong 2.6(1 - \beta)^{0.5}\exp(-3\beta), \alpha\gamma \cong \beta^{1.23}. \tag{16}$$

Here, τ_{HN} and τ_{KWW} are the relaxation times of Haviliak-Negami equation and KWW function, respectively.

2.2 Dynamic specific heat

In the equilibrium statistical mechanics, a specific heat at a constant pressure C_p is defined by

$$C_p = \frac{\partial \langle H \rangle}{\partial T}, \tag{17}$$

where H is enthalpy. In this definition, C_p is time-independent and includes the contribution of the total degrees of freedom.

According to the nonequilibrium statistical mechanics, dynamic (or complex) specific heat is $C_p(\omega)$ is defined by the fluctuation-dissipation theorem [16, 17] using the correlation function of enthalpy fluctuations. In the classical limit, it holds that

$$C_p(\omega) = C_p^0 + \frac{i\omega V}{k_B T^2}\int_0^\infty \langle [h(t) - \overline{h}][h(0) - \overline{h}]\rangle e^{i\omega t}dt, \tag{18}$$

where, $h(t)$ is the time-dependent enthalpy at constant pressure, $\bar{h}$ is the averaged $h(t)$, and V is the volume.

The specific heat of nonequilibrium materials is generally not a constant on time. The specific heat of nonequilibrium states has been discussed using a response function $\phi(t)$. The frequency-domain enthalpy is defined by the change of enthalpy $h(t) \equiv \delta H = H(t) - H_{eq}$, where H_{eq} is the value at an equilibrium state, is given by

$$h(\omega) = \int_0^\infty h(t)e^{i\omega t}dt = C_p(\omega)T(\omega) \tag{19}$$

The dynamic-specific heat $C_p(\omega)$ is related to the enthalpy response function $\phi(t)$.

$$C_p(\omega) = C_p^\infty + \left(C_p^0 - C_p^\infty\right)\int_0^\infty \phi(t)e^{i\omega t}dt, \tag{20}$$

where C_p^∞ is the value that equilibrates instantaneously, and C_p^0 is the static specific heat in an equilibrium state.

The specific-heat spectroscopy was developed by Birge [6], using an alternating current calorimetry technique, is not generally in use in laboratories and requires expert skill to detect weak third harmonic signals. In fact, this method measures the product of specific heat and thermal conductivity, i.e., the thermal effusivity [18]. Photoacoustic spectroscopy was also applied to study the dynamic specific heat of glass-forming materials [7]. Recently, temperature-modulated DSC (MDSC) is commercially available, and measurement can be performed in routine DSC scan [19]. In this chapter, at first, MDSC and the application to liquid-glass transitions of oxide network glasses are described. Secondly, photoacoustic spectroscopy is described, and the application to liquid-glass transitions of alcohols is described.

3. Temperature-modulated differential scanning calorimetry

In the early 1990s, a new type of differential scanning calorimetry (DSC), modulated temperature DSC (MDSC), was developed by Reading et al. [20, 21]. It has been shown that MDSC is a powerful tool to investigate the dynamical properties of glass transitions [22, 23]. MDSC is the extended equipment of a standard DSC. A small sinusoidal temperature perturbation is superimposed on the linear temperature ramping of the standard DSC, as shown in **Figure 6**. The MDSC was analyzed by Schawe based on the linear-response theory [24, 25].

The calorimeter response is given by

$$\frac{dQ}{dt} = C_p\frac{dT}{dt} + f(t, T). \tag{21}$$

Here, dQ/dt, dT/dt, C_p, and $f(t, T)$ are heat flow, heating rate, specific heat, and kinetic component, respectively. This heating rate-dependent component is called reversing, and the other independent one is nonreversing. MDSC can simultaneously measure these two components and the total DSC heat flow.

In DSC, the differential heat flow (HF) was determined by the temperature difference between the sample and the reference. The resultant HF is analyzed by the

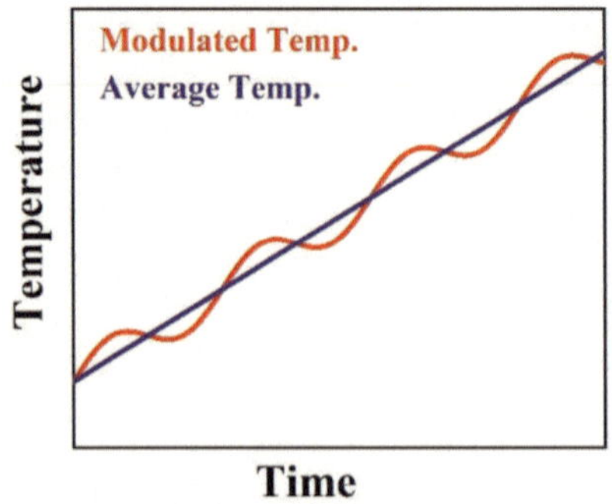

Figure 6.
The temperature-modulated DSC (MDSC) applies a small temperature fluctuation, generally sinusoidal temperature oscillation, to linear temperature ramping of conventional differential scanning calorimetry.

deconvolution of the sample's response into the underlying heating rate [26]. The raw resultant HF is averaged over the period of more than one modulation. After the averaged HF was subtracted from the resultant HF, the modulated contribution to the resultant HF is calculated by the discrete Fourier transformation (DFT) to obtain the amplitude of modulated contribution A_{HF}, and the phase lag (ϕ) between the modulated contribution and modulation in the heating rate. The phase angle is corrected according to the calculation reported in Ref. [27]. The absolute value of C_p^* (ω) is calculated by the amplitudes of modulated contribution A_{HF} and the heating rate A_q as shown in Eq. (22). The real part C_p' and imaginary part C_p'' are calculated by the phase lag(ϕ) from C_p^*:

$$\left|C_p^*\right| = \frac{A_{HF}}{MA_q}, \tag{22}$$

where M is the mass of the sample.

Modulated temperature differential scanning calorimetry (MDSC) measurements were performed by using TA instruments DSC2920 to obtain a thermal relaxation process in the vicinity of a glass transition temperature [26]. **Figure 7** shows the

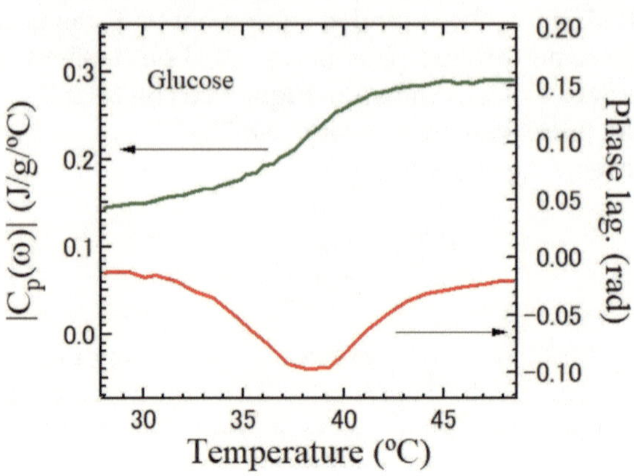

Figure 7.
The curves of the absolute value of complex specific heat |Cp| and phase lag of glucose obtained by MDSC [26].

absolute value of $C_p{}^*$ and phase lag(ϕ) observed in the vicinity of a glass transition temperature of glucose. It has been shown that MDSC can provide the frequency-dependence of the specific heat in the order of 0.01–0.1 Hz. The information obtained by MDSC determines the frequency-dependent specific heat $C_p{}^*(\omega)$. Then, $C_p'(\omega)$ and $C_p''(\omega)$ are calculated using a phase lag ϕ,

$$C_p^*(\omega) = \left|C_p^*(\omega)\right|e^{-\phi} = C_p'(\omega) - iC_p''(\omega) \tag{23}$$

$$C_p'(\omega) = \left|C_p^*(\omega)\right| \cos\phi, C_p''(\omega) = \left|C_p^*(\omega)\right| \sin\phi, \tag{24}$$

where $C_p'(\omega)$ relates to an in-phase component of the modulation and $C_p''(\omega)$ relates to an out-of-phase component. The temperature dependence of real C_p' and imaginary C_p'' parts of glass-forming glucose is shown in **Figure 8**.

The dynamic heat capacity of sodium borate glasses was measured by DSC 2920 and DSC T-zero Q200 (TA Instruments, Tokyo, Japan) [28]. A glass sample of about 10.0 mg was put into an aluminum pan. The sample pan was heated over T_g with a linear heating rate of 0.5 or 1.0°C/min. The temperature modulation conditions are amplitude ±1 °C and the period between 20 and 200 s. As a purge gas, dry nitrogen gas flowed in a sample chamber with a flow rate of 20 ml/min. The frequency dependences of the real and imaginary parts of the dynamic heat capacity of $x Na_2O$-(1 − x)B_2O_3 glass (x = 0.41) in the vicinity of glass transition temperature are shown in **Figure 9** [28].

In **Figure 9**, the peak temperature of $C_p''(\omega), T_g(\omega)$, increases as the temperature modulation frequency increases. The α-relaxation time τ_α is related to the modulation period, and it was determined by

$$\tau_\alpha = \frac{1}{\omega} = \frac{P}{2\pi}, \tag{25}$$

where P is the temperature modulation period. **Figure 10** shows the Angell plot [29] on the observed values of τ_α, where the horizontal axis is normalized by static T_g, which was determined as the temperature when τ_α becomes 100 s. By the Angell plot, fragility index m is determined by

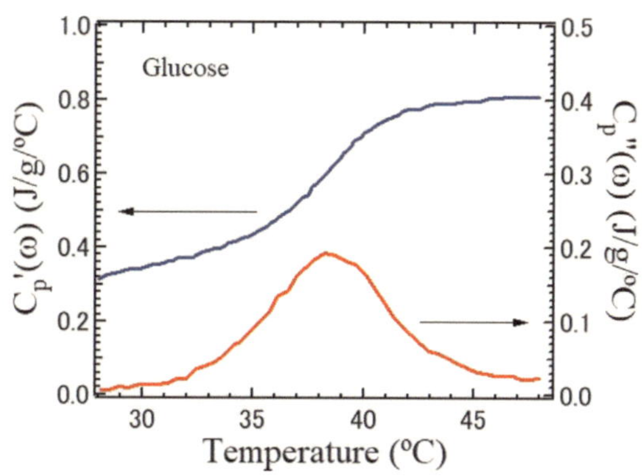

Figure 8.
The real and imaginary parts of dynamic specific heat in the vicinity of a glass transition temperature of glucose.

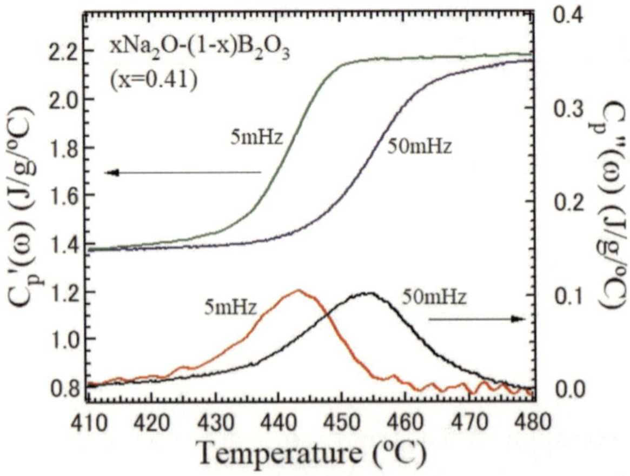

Figure 9.
Frequency dependences of the real and imaginary parts of the dynamic specific heat of xNa_2O-$(1 - x)B_2O_3$ *glass* ($x = 0.41$) *observed by T-zero DSCQ200.*

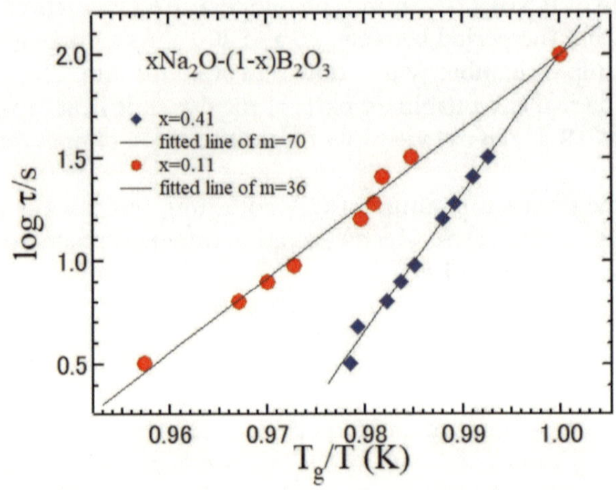

Figure 10.
Angell plot of xNa_2O-$(1 - x)B_2O_3$ *glasses with* $x = 0.11$ *and* 0.41 *Mol%. The horizontal axis is normalized by* T_g*. The straight lines are the fits by a least square method.*

$$m = \left[\frac{dlog\tau_\alpha}{d\left(\frac{T_g}{T}\right)}\right]_{T=T_g}. \tag{26}$$

The range of m changes from strong glass-forming liquids with $m \approx 16$ to fragile glass-forming liquids with $m \approx 200$ [30]. From the Angell plots of x = 11 and 41 mol% in **Figure 10**, the fragility index m was calculated by Eq. (26), corresponding to the steepness of the linear curve. Such a determination of fragility index using MDSC was also reported in lithium borate glasses [31], glucose, fructose [26], ethylene glycol-glucose aqueous solutions [32], and poly (propylene glycol)s [33].

The Na_2O content dependence of the fragility index m of xNa_2O-$(1 - x)B_2O_3$ glasses is shown in **Figure 11** with the reference values summarized by

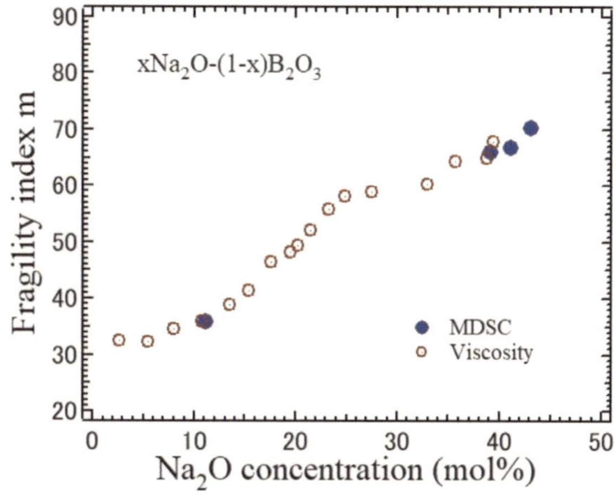

Figure 11.
Composition dependence of the fragility index m of xNa_2O*-*$(1 - x)B_2O_3$ *glasses. The values observed by the MDSC are plotted by solid circles. The reference values summarized by Chryssikos et al. [34] are plotted by open circles.*

Chryssikos et al. [34]. The values observed by the MDSC are in good agreement with the experimental result of viscosity. The agreement indicates the reliability of the MDSC in studying the fragility of glass-forming materials. It is found that the steepness increases as the Na_2O content increases. This result indicates the change of the fragility from "strong" to "fragile" as the Na_2O content increases below $x = 0.45$. According to the Vilgis model, the degree of fragility relates to the distribution of the coordination number of a key element. In the sodium borate glasses, the composition dependence of m and the variation of the coordination number of boron atoms have similarities. It indicates that the origin of the fragility can be related to the distribution of the coordination number of boron. A similar result was also reported in lithium borate glass using MDSC [31].

For the study on the non-Debye nature of structural relaxation, the Cole-Col plot of xLi_2O-$(1 - x)B_2O_3$ glass ($x = 0.40$) is shown in **Figure 12**. The imaginary part, C_p'' is plotted against the real part C_p'. Experimental conditions: heating rate 1°C/min, amplitude of modulation 1.0°C, period of modulation 100 s. This plot shows the

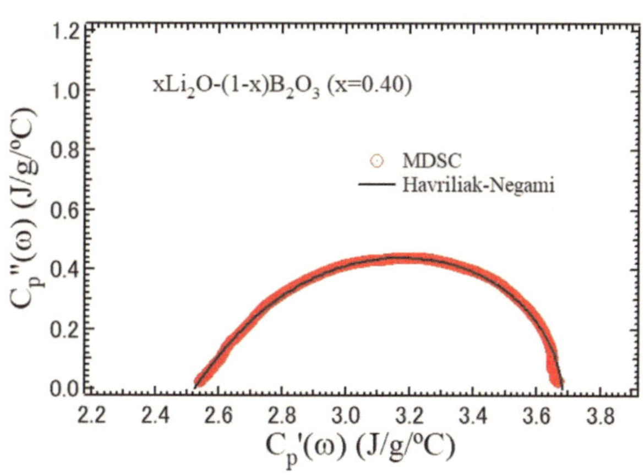

Figure 12.
Cole-Cole plot of dynamic specific heat of 40 Li_2O*·60* B_2O_3*. Circles denote the experimental results. The solid line denotes the fitted values by Haviliak-Negami equation (*$\alpha = 0.94$ *and* $\gamma = 0.67$*).*

temperature averaged non-Debye nature in the vicinity of a dynamic glass transition temperature. Circles denote the experimental results. The solid line denotes the fitted values by the HN equation. The values of parameters are $\alpha = 0.94$ and $\gamma = 0.67$, and both parameters are less than 1.0, and non-Debye nature is observed [17].

The origin of non-Debye nature relates to the coexistence of the different types of intermediate structural units with the three- and four-coordinated boron atoms. In alkali borate glasses, the addition of Li_2O leads to the formation of tetrahedral $B\varnothing_4$, where $\varnothing$ is bridging oxygen. Therefore, the heterogeneity of the intermediate structure occurs under a low composition range of Li_2O. The distribution of intermediate structures causes the distribution of relaxation times and the non-Debye nature. Such a non-Debye nature was also observed by the Cole-Cole plot of complex heat capacity in sodium borate glass [28], fructose, and glucose [26].

4. Photoacoustic method

In 1880, the photophonic phenomena were discovered by Alexander Graham Bell. When a solid in an enclosed cell is illuminated by a rapidly modulated beam of sunlight shines, it is possible to hear an audible sound by a hearing tube attached to the cell [35]. Tyndall and Rontgen reported that an audible sound could also be emitted when a gas in an enclosed cell is illuminated with intermitted Drummond's limelight [36, 37]. Bell subsequently experimented with the production of sound by radiant energy with a variety of solids, liquids, and gases [38]. After 100 years had passed, the interest in photoacoustic effects and related experimental techniques increased. The first conference on photoacoustic effect took place in February 1981 in Bad Honnef in Germany [39]. Spectroscopy and detection of minute concentrations, monitoring chemical reactions, energy conversion processes, and calorimetric applications to phase transitions were extensively discussed. By using a finely focused laser beam, photoacoustic imaging (PAI) is also possible, and a liquid-solid interface was observed in the vicinity of a melting temperature [40]. Recently, PAI is an emerging noninvasive imaging modality. PAI provides better resolution than pure ultrasonic imaging and deeper penetration than pure optical imaging [41].

Rosencwaig and Gersho reported a quantitative derivation for the acoustic signal of solid and semisolid matter in a photoacoustic cell in terms of the optical, thermal, and geometric parameters of the system [42]. A one-dimensional model of the heat flow in the cell resulting from the absorbed light energy was calculated.

A standard cylindrical photoacoustic cell is shown in **Figure 13** by the cross-sectional view. A sample is located between backing material and gas. Here, the thermal diffusion length of a sample, μ_s, the thermal diffusion length of gas, μ_g, the thermal conductivity (cal/cm s °C), k_i, the density (g/cm^3), ρ_i, the specific heat (cal/g° C), C_{pi}, the thermal diffusivity i (cm^2/s) $\alpha_i = k_i/\rho_i C_i$, the thermal diffusion coefficient (cm), $a_i = (\omega/2\alpha_i)^{1/2}$; the thermal diffusion length (cm) $\mu_i = 1/a_i$, where subscript i denotes s, g, and b for the solid, gas, and backing material, respectively. The modulated frequency of the incident light to a cell is ω in radians per second.

The thermal diffusion equation in the solid, taking into account the distributed heat source, can be written by

$$\frac{\partial^2 \phi}{\partial x^2} = \frac{1}{\alpha_s}\frac{\partial \phi}{\partial t} - Aexp(\beta x)[1 + \exp(j\omega t)], \quad \text{for} - L_s \leq x \leq 0. \tag{27}$$

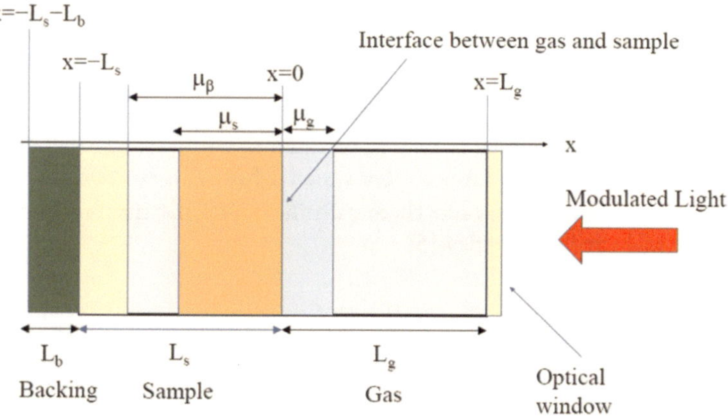

Figure 13.
One-dimensional model of photoacoustic effect in gas-microphone method on the optically thick sample ($L_s > \mu_\beta > \mu_s$), where μ_s and μ_g are thermal diffusion lengths of a sample and gas, respectively. $\mu_\beta = 1/\beta$ is the optical absorption length.

Here, $A = \beta I_0 \eta / 2k_s$, ϕ and η are temperature and efficiency, respectively. The full expression for the incremental pressure,

$$\delta P(t) = Q \exp\{j(\omega t - \pi/4)\}, \text{where } Q = \gamma P_0 \theta / \sqrt{2} l_g a_g T_0. \tag{28}$$

The full expression for $\delta P(t)$ is complicated for interpretation. However, it is understandable by considering typical conditions. For optically opaque ($\mu_\beta = 1/\beta \ll 1$) and thermally thick ($\mu_s \ll l$; $\mu_\beta < \mu_s$) solids, it holds that

$$Q \cong \frac{(1-j)\mu_s}{2a_g}\left(\frac{\mu_s}{k_s}\right)Y, \text{where } Y = \frac{\gamma P_0 I_0}{2\sqrt{2} l_g T_0}. \tag{29}$$

This case is shown in **Figure 14(a)** [42, 43].

The photoacoustic signals of glass-forming liquids were measured in the condition of **Figure 14(b)** [7, 44]. The modulated light is absorbed in an opaque film, and

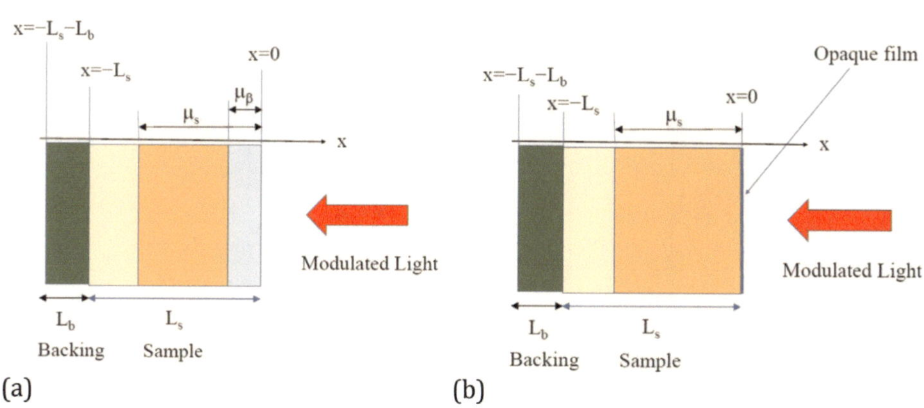

Figure 14.
(a) Optically thin and thermally thick ($\mu_\beta < \mu_s < L_s$), (b) optically very thin and thermal thick (μ_β $\mu_s < L_s$). The modulated light is absorbed in an opaque film, and thermal waves are generated at interfaces with adjacent gas and a sample.

thermal waves are generated at interfaces with adjacent gas ($0 < x < L_g$) and a sample ($-L_s < x < 0$). During a cycle of modulation, thermal waves propagate μ_s and μ_g to minus and plus directions along the x-axis, respectively.

According to their theory on the case of an optically very thin and thermal thick sample, the photoacoustic signal is related to squared complex thermal effusivity, $C_p(\nu)\kappa(\nu)$,which is the product of complex specific heat and complex thermal conductivity, as below. Linear-response theory defines dynamic thermal conductivity by the fluctuation of heat flux density [5]:

$$|C_p(\nu)\kappa(\nu)| = A_{pt}(\nu)^{-2}. \tag{30}$$

$$tan^{-1}\left\{\frac{(C_p(\nu)\kappa(\nu))''}{(C_p(\nu)\kappa(\nu))'}\right\} = -2\phi_{pt}(\nu). \tag{31}$$

Here, $A_{pt}(\nu)$ and $\phi_{pt}(\nu)$ are the frequency-dependent photoacoustic amplitude and the phase lag, respectively. The $C_p(\nu)$ and $\kappa(\nu)$ are dynamic specific heat and dynamic thermal conductivity, respectively. From Eqs. (30) and (31), the frequency-dependence of the product of a dynamic specific heat and a dynamic thermal conductivity can be determined by the frequency-dependence of observed $A_{pt}(\nu)$ and $\phi_{pt}(\nu)$. **Figure 15** shows the temperature dependence of A_{pt} and $-2\ \phi_{pt}$ of propylene glycol at 80 Hz [44]. Upon cooling from a high temperature, the relaxation frequency (the inverse of the relaxation time) decreases and approaches the modulation frequency of an incident light. When the relaxation frequency becomes the same as the modulation frequency, the maximum of $-2\ \phi_{pt}$ appears and A_{pt} decreases.

Cole-Cole plot of dynamic $C_p\kappa$ of propylene glycol is shown in **Figure 16** [44]. These values of the Davidson-Cole exponent of photoacoustic method were compared with the values determined by the dielectric and the ultrasonic measurements, as shown in **Table 1**. It is found that the Davidson-Cole exponent of photoacoustic method in the two liquids is smaller than those of the dielectric and ultrasonic measurements [45, 46]. The dielectric relaxation is related to only the fluctuation of polarization. In contrast, a photoacoustic response is related to thermal fluctuations,

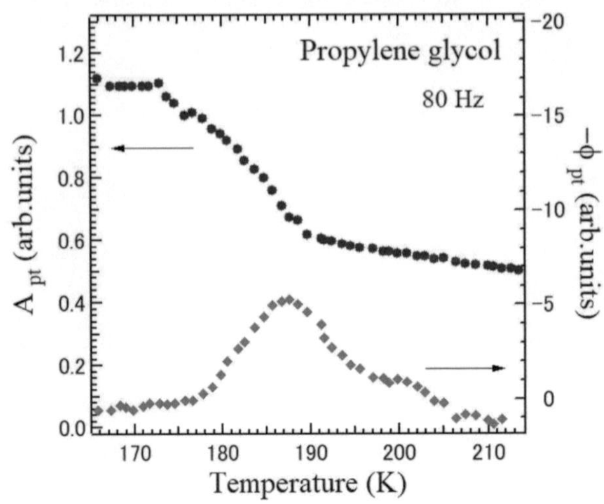

Figure 15.
Temperature dependence of amplitude and phase of photoacoustic signals in propylene glycol [44].

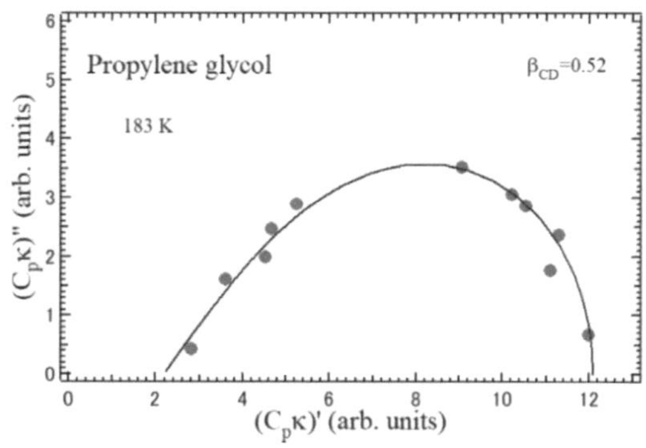

Figure 16.
Cole-Cole plot of dynamic $C_p\kappa$ of propylene glycol. The solid line shows calculated values by the Davidson-Cole equation at β_{CD}=0.52, where $(C_p\kappa)'$ and $(C_p\kappa)''$ are real and imaginary parts of dynamic $C_p\kappa$ [44].

Method	Response	Glycerol	Propylene glycol
Dielectric [45]	Polarization	0.58 ± 0.03	0.66 ± 0.01
Ultrasonic [46]	Strain	0.42 ± 0.05	
Photoacoustic [44]	Enthalpy	0.35 ± 0.03	0.52 ± 0.05

Table 1.
Comparison of Davidson-Cole exponents of dielectric, ultrasonic, and photoacoustic measurements in glass-forming glycerol and propylene glycol.

which contain all degrees of freedom including polarization and strain. Therefore, the distribution of relaxation time of photoacoustic method can be broader than those of the dielectric and ultrasonic measurements. Therefore, it is reasonable that Davidson-Cole exponents of photoacoustic method can be smaller than those of the dielectric and elastic measurements. This fact is seen in **Table 1**. There are two and three hydroxyl groups in a molecule for propylene glycol and glycerol, respectively. Since the dominant interaction among molecules is the intermolecular hydrogen bonding, the nearest interaction of molecules of glycerol is more complex than that of propylene glycol, and the distribution of relaxation time of glycerol can be broader than that of propylene glycol. Therefore, the exponent of glycerol is smaller than that of propylene glycol, which is smaller than those of n- and i-propanol [47].

Upon cooling a liquid from a high temperature, several kinds of relaxation processes, an α-relaxation, slow and fast β-processes and the low-energy excitation, boson peaks were observed in the large frequency range between the μHz and THz ranges as shown in **Figure 17** [9, 48, 49]. The relaxation time of the main α-structural relaxation process in a fragile liquid diverges at T_{VF}, and it obeys the following Vogel-Tammann-Fulcher law reflecting the cooperative motion in a cage [2].

$$\tau = \tau_0\, exp\left(\frac{B}{T - T_{VF}}\right). \tag{32}$$

Here τ_0 and B are constants. However, in a strong liquid, such as a silica glass, T_{VF} is far below T_g, and the relaxation time approximately obeys the Arrhenius law. Therefore, no remarkable change in relaxation time is observed in the vicinity of T_g.

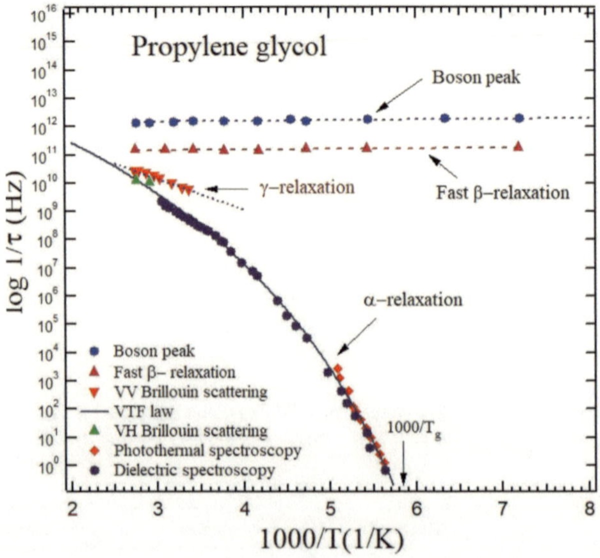

Figure 17.
Various relaxation processes and boson peaks of propylene glycol. The solid line shows the Vogel-Tammann-Fulcher (VTF) law observed by various measurements.

Propylene glycol undergoes a liquid-glass transition at about T_g = 168 K. It is one of the typical glass-forming materials with the strong glass-forming tendency [48, 49]. It belongs to an intermediate liquid with a fragility index of m = 48 in the strong-fragile classification [33]. The temperature dependences of α relaxation determined by dielectric and photoacoustic measurements together with the boson peak and fast relaxations determined by the Raman and Brillouin scattering measurements are shown in **Figure** 7. The boson peak frequency is nearly constant. The fast relaxation determined from the polarized Brillouin scattering spectra is also temperature-independent. In the supercooled liquid, this process is faster than other relaxation processes determined by dielectric and photoacoustic measurements [9, 44]. This process may be attributed to the fast β-process predicted by the mode coupling theory on glass transitions, which is faster than the α-relaxation process [50]. Another relaxation process was determined from the Rayleigh wing in a depolarized Brillouin scattering spectrum, and it was attributed to the α-relaxation at a GHz range [49]. The γ-relaxation was also determined from the width of a longitudinal acoustic (LA) peak, assuming the coupling between the relaxation process and LA mode [51]. Below a GHz range, the relaxation time of the α-relaxation process was observed by dielectric spectroscopy, which detects polar motions [9]. While dynamic heat capacity detects both polar and nonpolar motions. The mean relaxation time of the α-relaxation process obeys the Vogel-Tammann-Fulcher law. For the detailed study on the relaxation processes and vibrational properties in a liquid-glass transition, a complete picture of relaxation map is required in the very large frequency range, in which various kinds of relaxation processes and low-energy excitation are included.

5. Conclusions

In a linear-response theory, the dynamic specific heat is defined by Kubo's formula on the fluctuation-dissipation theorem using the correlation function of enthalpy fluctuations. It is important that the dynamic specific heat can provide us with a more general interpretation of various relaxation and dynamical processes. In fact, the origin of the enthalpy relaxation is attributed to the total degrees of freedom of atomic motions. While, the dielectric relaxation is related only to polar atomic motions, and ultrasonic spectroscopy is related only to strain. This chapter introduces two experimental methods to measure physical properties related to dynamic specific heat, (1) temperature-modulated differential calorimetry (MDSC) and (2) photoacoustic spectroscopy. Then, the dynamical properties of glass transitions in oxide network glasses and hydrogen-bonded alcohols are reviewed on dynamic glass transition temperature, fragility index, and the distribution of relaxation time.

In contrast, in broadband dielectric spectroscopy, which covers from micro- to terahertz frequency range, the frequency range of observable dynamic heat capacity is limited in the low-frequency range. MDSC detects dynamic specific heat, while specific-heat spectroscopy and photoacoustic methods are based on thermal effusion [18] and detect dynamic thermal effusion, which relates to the product of dynamic specific heat and dynamic thermal conductivity. Therefore, further development of experimental methods is necessary to extend the frequency range of measurements of dynamic specific heat or to detect dynamic specific heat and dynamic thermal conductivity independently.

Author details

Seiji Kojima
Division of Materials Science, University of Tsukuba, Tsukuba, Ibaraki, Japan

*Address all correspondence to: kojima@ims.tsukuba.ac.jp

References

[1] Kauzmann W. The nature of the glassy state and the behavior of liquids at low temperatures. Chemical Reviews. 1948;**43**:219-256. DOI: 10.1021/cr60135a002

[2] Fulcher GS. Analysis of recent measurements of the viscosity of glasses. Journal of the American Ceramic Society. 1925;**8**:339-355. DOI: 10.1111/j.1151-2916.1925.tb16731.x

[3] Lunkenheimer P, Loidl A. Dielectric spectroscopy of glass-forming materials: α-relaxation and excess wing. Chemical Physics. 2002;**284**:205-219. DOI: 10.1016/S0301-0104(02)00549-9

[4] Kojima S. Anniversary of Brillouin scattering: Impact on materials science. Materials. 2022;**15**:3518. DOI: 10.3390/ma15103518

[5] Kubo R. Statistical-mechanical theory of irreversible processes. I. General theory and simple applications to magnetic and conduction problems. Journal of Physical Society of Japan. 1957;**12**(6):570-586. DOI: 10.1143/JPSJ.12.570

[6] Birge NO, Nagel SR. Specific heat spectroscopy of the glass transition. Physical Review Letters. 1985;**54**(25):2674-2677. DOI: 10.1103/physrevlett.54.2674

[7] Kojima S, Kashiwada T, Nakahira S. Photoacoustic measurements of thermal dispersion in glass-transitions of glycerol and propylene glycol. High Temperature-High Pressures. 1991;**23**:701-704

[8] Debye P. Zur Theorie der anomalen Dispersion im Gebiete der langwelligen elektrischen Strahlung. Verh Deut Phys Gesell. 1913;**15**:777-793

[9] Park IS, Saruta K, Kojima S. Dielectric relaxation and calorimetric measurements of glass transition in the glass-forming dihydroxy alcohols. Journal of Physical Society of Japan. 1998;**67**(12):4131-4138. DOI: 10.1143/JPSJ.67.4131

[10] Davidson DW, Cole RH. Dielectric relaxation in glycerol, propylene glycol, and n-propanol. Journal of Chemical Physics. 1951;**19**(12):1484-1490. DOI: 10.1063/1.1748105

[11] Cole KS, Cole RH. Dispersion and absorption in dielectrics, I. Alternating current characteristics. Journal of Chemical Physics. 1941;**9**(4):341-351. DOI: 10.1063/1.1750906

[12] Havriliak S, Negami S. A complex plane analysis of α-dispersions in some polymer systems. Journal of Polymer Science Part. 1996;**14**:99-117

[13] Kohlrausch R. Theorie des elektrischen rückstandes in der Leidener flasche. Annalen der Physik. 1854;**167**:161-320. DOI: 10.1002/andp.18541670203

[14] Willams G, Watts DC. Non-symmetrical dielectric relaxation behavior arising from a simple empirical decay function. Transaction of the Faraday Society. 1970;**66**:80-85. DOI: 10.1039/TF9706600080

[15] Alvarez F, Alegria A, Colmenero J. Relationship between the time-domain Kohlrausch-Williams-Watts and frequency-domain Havriliak-Negami relaxation functions. Physical Review B. 1991;**44**(14):7306-7312. DOI: 10.1103/PhysRevB.44.7306

[16] Birge NO. Specific-heat spectroscopy of glycerol and propylene glycol near the

glass transition. Physical Review B. 1986; **34**(3):1631-1642. DOI: 10.1103/PhysRevB.34.1631

[17] Matsuda Y, Fukawa Y, Ike Y, Kodama M, Kojima S. Dynamic specific heat, glass transition and non-Debye nature of thermal relaxation in Lithium borate glasses. Journal of the Physical Society of Japan. 2008; **77**(8):084602. DOI: 10.1143/JPSJ.77.084602

[18] Christensen T, Olsen NB, Dyre JC. Conventional methods fail to measure cp (ω) of glass-forming liquids. Physical Review E. 2007;**75**(4):041502. DOI: 10.1103/PhysRevE.75.041502

[19] Matsuda Y, Matsui C, Ike Y, Kodama M, Kojima S. Non-Debye nature in thermal relaxation and thermal properties of Lithium borate glasses studied by modulated DSC. Journal of Thermal Analysis and Calorimetry. 2006;**85**(3):725-730. DOI: 10.1007/s10973-006-7652-9

[20] Reading M, Elliott D, Hill VL. A new approach to the calorimetric investigation of physical and chemical transitions. Journal of Thermal Analysis. 1993;**40**:949-955. DOI: 10.1007/BF02546854

[21] Reading M, Luget A, Wilson R. Modulated differential scanning calorimetry. Thermochimica Acta. 1994; **238**:295-307. DOI: 10.1016/S0040-6031(94)85215-4

[22] Park I, Kojima S. Glass transition and structural relaxation of intermediate liquids by MDSC and dielectric spectroscopy. Thermochimica Acta. 2000;**352-353**:147-152. DOI: 10.1016/S0040-6031(99)00459-1

[23] Carpentier L, Bustin O, Decamps K. Temperature-modulated differential scanning calorimetry as a specific heat spectroscopy. Journal of Physics D. 2002; **35**(4):402-408. DOI: 10.1088/0022-3727/35/4/317

[24] Schawe JEK. Principles for the interpretation of modulated temperature DSC measurements. Part 1. Glass transition. Thermochimica Acta. 1995; **261**(1):183-194. DOI: 10.1016/0040-6031(95)02315-S

[25] Schawe JEK, Höhne GWH. The analysis of temperature modulated DSC measurements by means of the linear response theory. Thermochimica Acta. 1996;**287**(2):213-223. DOI: 10.1016/0040-6031(96)88984-2

[26] Ike Y, Seshimo Y, Kojima S. Complex heat capacity of non-Debye process in glassy glucose and fructose. Fluid Phase Equilibria. 2007;**256**(1–2):123-126. DOI: 10.1016/j.fluid.2007.04.017

[27] Weyer S, Hensel A, Schick C. Phase angle correction for TMDSC in the glass-transition region. Thermochimica Acta. 1997;**304**(305):267-275. DOI: 10.1016/S0040-6031(97)00180-9

[28] Fukawa Y, Matsuda Y, Kawashima M, Kojima S. Determination of complex-specific heat and fragility of sodium borate glasses by temperature-modulated DSC. Journal of Thermal Analysis and Calorimetry. 2010;**99**: 39-44. DOI: 10.1007/s10973-009-0522-5

[29] Angell CA, MacFarlane DR, Oguni M. The Kauzmann paradox, metastable liquids, and ideal glasses: A summary. Annals of the New York Academy of Sciences. 1986;**484**(1): 241-247. DOI: 10.1111/j.1749-6632.1986.tb49574.x

[30] Vashist P, Patial BS, Bhardwaj S, Tripathi SK, Thakur N. On the non-isothermal crystallization kinetics, glass

forming ability and thermal stability of Bi additive Se–Te–Ge alloys. Journal of Thermal Analysis and Calorimetry. 2023; **148**(15):7717-7726. DOI: 10.1007/s10973-023-12271-5

[31] Matsuda Y, Fukawa Y, Kawashima M, Mamiya S, Kojima S. Dynamic glass transition and fragility of lithium borate binary glasses. Solid State Ionics. 2008;**179**(40):2424-2427. DOI: 10.1016/j.ssi.2008.09.011

[32] Hashimoto E, Seshimo Y, Sasanuma K, Aoki Y, Kanazawa H, Ike Y, et al. Dynamical properties of ethylene glycol and glucose studied by temperature-modulated differential scanning calorimetry and Brillouin scattering. Journal of Thermal Analysis Calorimetry. 2010;**99**(1):45-50. DOI: 10.1007/s10973-009-0543-0

[33] Koda S, Shibata T, Park IS, Kojima S. Relaxation dynamics and fragility during liquid-glass transitions of poly (propylene glycol)s. Current Applied Physics. 2015;**15**(7):805-810. DOI: 10.1016/j.cap.2015.04.041

[34] Chryssikos GD, Kamitsos EI, Yiannopoulos YD. Towards a structural interpretation of fragility and decoupling trends in borate systems. Journal of Non-Crystalline Solids. 1996;**196**:244-248. DOI: 10.1016/0022-3093(95)00594-3

[35] Bell AG. On the production and reproduction of sound by light. American Journal of Science. 1880;**s3-20(118)**: 305-324. DOI: 10.2475/ajs.s3-20.118.305

[36] Tyndall J. Action of an intermittent beam of radiation heat upon gaseous matter. Proceedings of Royal Society of London. 1881;**31**:307-317. DOI: 10.1098/rspl.1880.0037

[37] Rontgen WC. On tones produced by the intermittent irradiation of a gas. Philosophical Magazine Series 5. 1881; **11**(68):308-311. DOI: 10.1080/14786448108627021

[38] Bell AG, LXVIII. Upon the production of sound by radiant energy. Philosophical Magazine Series 5. 1881; **11**(71):510-528. DOI: 10.1080/14786448108627053

[39] Luscher E, Coufal HJ, Korpiun P, Tilgner R, editors. Photoacoustic Effect. Braunschweig: Friedr. Vieweg & Sohn; 1984

[40] Kojima S. Photoacoutic imaging for the melting of gallium. Japanese Journal of Applied Physics. 1984;**24**(12R): 1571-1572. DOI: DOI 10.1143/JJAP.24.1726

[41] Ghaliev R. Photoacoustic Imaging: Principles. IntechOpen, London, UK: Advances and Applications; 2020

[42] Rosencwaig A, Gersho A. Theory of the photoacoustic effect with solids. Journal of Applied Physics. 1976;**47**: 64-69. DOI: 10.1063/1.322296

[43] Kojima S. Study of phase transitions by frequency dependent photoacoustic measurements. In: Leroy O, Breazeal MA, editors. Physical Acoustics. New York: Plenum Press; 1991. pp. 399-404

[44] Kojima S, Koketsu T, Takahashi E, Kanayasu M. Complex thermal effusivity of glassy polyhydric alcohols. Fluid Phase Equilibria. 1993;**88**: 209-218. DOI: 10.1016/0378-3812(93)87112-E

[45] Angle CA, Smith DL. Test of the entropy basis of the Vogel-Tammann-Fulcher equation. Dielectric relaxation of polyalcohols near T_g. Journal of Physical Chemistry. 1982;**86**(19):3845-3852. DOI: 10.1021/j100216a028

[46] Jeong YH, Nagel SR, Bhattachya S. Ultrasonic investigation of the glass transition in glycerol. Physical Review A. 1986;**34**(1):602-608. DOI: 10.1103/PhysRevA.34.602

[47] Park IS, Mikami M, Kojima S, Takagi Y. Dynamics of liquid-glass transition in propanol. AIP Conference Proceedings. 1999;**469**:525-526. DOI: 10.1063/1.58534

[48] Kojima S, Novikov V. Correlation of temperature dependence of Quasielastic light scattering intensity and alpha-relaxation time. Physical Review B. 1996; **54**(1):222-227. DOI: 10.1103/PhysRevB.54.222

[49] Kojima S, Sato H, Yoshihara A. Light scattering of supercooled propylene glycol. Journal of Physics: Condensed Matter. 1997;**9**(46):10079-10086. DOI: 10.1088/0953-8984/9/46/005

[50] Gotze W, Sjogren L. Relaxation processes in supercooled liquids. Reports on Progress in Physics. 1992;**55**(3): 241-376. DOI: 10.1088/0034-4885/55/3/001

[51] Takanashi K, Yoshihara A, Kojima S. Brillouin scattering study on dynamical properties of a viscoelastic liquid propylene glycol using a tandem Fabry-Perot interferometer. Japanese Journal of Applied Physics. 1994;**33**(5S):2867-2873. DOI: 10.1143/JJAP.33.2867